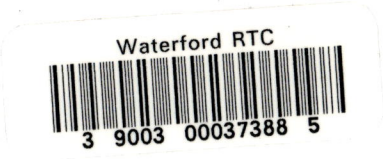

First Year Engineering
Information Technology and Systems

before the last d

First Year Engineering

Information Technology and Systems

D C Green

MTech, CEng, MIEE

Longman
Scientific &
Technical

Longman Scientific & Technical
Longman Group Limited
Longman House, Burnt Mill, Harlow
Essex, CM20 2JE, England
and Associated Companies throughout the world

Trademarks
Throughout this book trademarked names are used. Rather than put a
trademark symbol with every occurrence of a trademarked name, we
state that we are using the names only in an editorial fashion and to the
benefit of the trademark owner with no intention of infringement of the
trademark.

First published 1994

British Library Cataloguing in Publication Data
A catalogue entry for this title is available from the British Library

ISBN 0-582-21864-0

Set by 4 in Compugraphic 10/12 pt Times
Produced through Longman Malaysia, VVP

Contents

Preface

This book provides a comprehensive introduction to the three topics of utmost importance to all students starting on the first year of a course leading to qualification as a technician, and may also be useful during further studies as an engineer. The book covers the basic concepts of information technology, microelectronic systems, and communication systems and includes all the content of the BTEC First units on these topics.

The first five chapters cover the contents of the Information Technology unit 2864B and give an introduction to Information Technology (IT) and its various applications, before going on to consider the basic principles of operation of the digital computer and the applications of IT in the modern office. Chapters 6, 7 and 8 satisfy the requirements of the Microelectronic Systems unit 2869B, covering in turn analogue and digital signals, microelectronic systems, and an introduction to assembly language programming. The remaining chapters of the book are concerned with telecommunications and describe the basic principles of telephone networks, data communication, radio systems and television. These chapters cover the requirements of the BTEC Communication Systems unit 2872B.

D. C. G.

1 Information technology

Information comprises pieces of knowledge. Information is received from books, newspapers, magazines, and other printed material, computer systems, radio and television, and many other sources. The information used by a business may be purely numerical, graphical (graphs and charts), textual, or a mixture of any two, or of all three. The information could be, for example, records of the stocks of goods held, of goods ordered and not yet delivered, of invoices sent out to customers, of bills paid, and so on. A business would also have information on the wages and salaries of staff, their tax and National Insurance numbers, and their home addresses. If a domestic radio receiver is switched on information is received by the listener; this information may be about events that have occurred anywhere in the world, about the weather, about forthcoming programs on the radio, and so on. The information obtained from the radio may be purely numerical data, such as the time, a mixture of numerical and textual data, such as a weather forecast or the football results, or solely textual data, for example, a report on yesterday's events in Parliament. Should the same information be received from a television receiver some of it may, in addition, be presented using a graphical format.

In industry and commerce it is not enough merely to keep records; the stored information must be processed in some way and the results made readily available to anyone who needs to take some kind of action based upon that information. For example, a departmental manager may need to analyse sales or production figures to ascertain whether things are going well or there is a need to take action. **Information technology** (IT) is concerned with the handling of information. It involves the collection or capture of information, its storage and its later retrieval, its processing, and then, finally, the communication of the processed information to the point at which it will be used.

The technology employed in modern systems makes full use of the **digital computer** and the **microprocessor**. Both these electronic devices handle information in the form of **binary digits** or **bits**. This means that all information must first be converted into bits before it can be stored and processed by either a computer or a microprocessor. A bit may be either binary 0 or binary 1 and a combination of eight bits, known as a **byte**, represents one character in the **American Standard Code For Information Interchange** (ASCII) system. Some examples of the use of the ASCII code are given in Fig. 1.1.

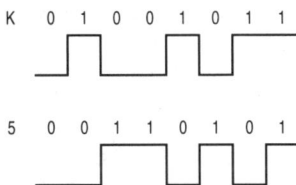

Fig 1.1 ASCII code for (a) the letter K and (b) the figure 5.

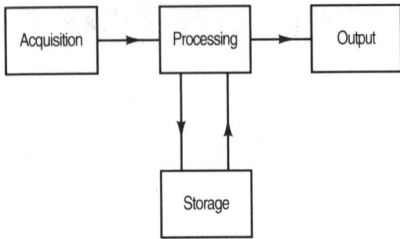

Fig. 1.2 Information technology system.

The basic block diagram of an information technology system is shown by Fig. 1.2, in which information is collected, processed and stored and, when required, can be retrieved and outputted. The information that is to be acquired may be in any one of several different forms, such as speech, music, video, data and text. Consider the production of a CD disk. The live music is first captured in the recording studio, or on stage, by one or more microphones. The sound is processed by sound mixers and other recording equipment to obtain the desired balance and to incorporate any special effects, such as echos. The processed sound is then converted from its original analogue form into the equivalent digital form by an **analogue-to-digital converter**. The digitalized music is then stored on a master disk. The master disk may be used at any time to retrieve the music and to copy it onto many other disks. The copied disks may then be distributed to record shops and department stores and the buyers can output the music by playing the disk on a CD player.

As another example of IT consider the receipt of a sales order written on paper in an office. The information may be stored in a folder, which in turn will probably be stored in a filing cabinet, or it may be entered into a computer system. The information can be retrieved by taking the order out of the folder, or by accessing the computer store, and can then be processed by determining how the order can be fulfilled, by the calculation of costs, the adjustment of sales records, etc. The processed information is then communicated; a sales invoice is sent to the customer, instructions are sent to the warehouse to deliver the goods, and the sale is included in the monthly sales figures that are later sent to the management. Monthly, or perhaps weekly, sales and stock figures are required by a business so that it is able to keep itself informed of the growth or decline of sales and the stocks held of various items, and is better able to keep control. It may often be necessary for the sales figure to be broken down into particular products, particular salespeople, different areas of the country, and so on, and the data may be given in the form of figures and/or graphs and pie charts. All of this is much easier and faster to carry out using electronic methods instead of the traditional paper-shifting.

Quality of information

The term 'quality of information' means that the information, of whatever kind, is accurate, timely and relevant, and is presented in a way that the user is able to understand.

Accuracy

If information is accurate it is without error: written information has no spelling or printing mistakes, tabulated figures are entered correctly and calculations have been carried out without a mistake, in a sales office the correct item(s) have been sent to the customer, and so on.

The computers and other digital electronic equipment that are employed to process information work at an extremely high speed and without error. Unfortunately speech, music, and other **analogue** signals cannot be converted into digital form without the introduction of some noise. A digital processing system is designed so that this noise does not have a noticeable effect upon the information being processed and stored. A CD player, for example, uses digital techniques and gives better quality reproduction of music than the older systems, such as record and cassette players, which employ analogue circuitry throughout. Thus, the quality of information obtained from a CD player is better than the quality of information obtained from a record player.

For textual information the use of a **word processor** and its incorporated spell checker will allow printed material to be produced more quickly and more accurately than if a typewriter is employed. The calculation of charts, graphs, results, statistics, etc., can be carried out more rapidly and with greater accuracy using a computer than would ever be possible manually.

Timeliness

Information should be sufficiently up-to-date for its intended usage. As newspapers are published daily, their topical information is, at most, only one day old; and as most magazines are published either weekly or monthly, their information may be a week or a month old; but the information provided by a book cannot possibly be as recent since the publication of a book usually takes about one year from the receipt of manuscript. Calculations on data can be carried out very quickly and accurately using a computer so that the information supplied to management can be more reliable and more up-to-date. In the scientific and engineering fields research information that is more than a few years old may no longer be relevant, while in the finance world information that is only a few days old may already be out-of-date. On the other hand, historical information never becomes out-of-date.

Relevance

The information stored in a modern information system can be sorted and evaluated before the relevant information is retrieved as, and when, it is wanted. All non-useful information can then be ignored. This makes it easier for the information relevant to a specific problem to be studied and decisions based upon that information to be made.

Information stored in a **database** (p. 111) can be transmitted over a telephone line to an enquirer and displayed upon the screen of a visual display unit (VDU) or a computer monitor. The relevance, or otherwise, of the information can then be determined and printed out, or loaded onto a disk, if it is required.

Presentation

Information should be presented in a form that can easily be understood and used by its recipient. Naturally, the 'best' method of presentation depends upon the application. A modern information technology system allows numerical information to be readily presented in the form of graphs and/or charts — instead of long lists of figures — and these graphs may be either printed out or displayed on a VDU or a monitor.

Message transmission

The information received by a recipient has often been gathered at some other point and has then been transmitted to the recipient. The weather forecasts that are broadcast by the radio and television authorities have been produced from data that has been gathered by the MET office. This data must be transmitted from the weather forecast studio to the radio or television receiver in the home. The weather information is collected by the weather forecasters, processed into actual weather forecasts, stored until the time for the weather forecast arrives, and then transmitted by radio wave, to the home radio or television receivers. This is one example of **message transmission**. Of course, the radio or television receiver itself forms another information system. The radio/television signal picked up by the receiving aerial is the information acquired by the system. This information is then processed to select the required signal and reject all others, to amplify the selected signal, and to shift its frequency from the incoming radio frequency to the audio-frequency range over which the human ear is able to hear.

In all cases there are four main parts to an information technology system: (i) information input, (ii) information processing, (iii) information storage, and (iv) information output in the form required by the user.

Spoken messages are commonly transmitted over the **public switched telephone network** (PSTN), and data messages are transmitted over either the PSTN, the **public switched data network** (PSDN) or via various private data networks. The telephone transmitter and receiver are both examples of **transducers** that are used to convert sound waves into electrical signals or vice versa. The PSTN is considered in Chapter 10, but essentially a telephone is connected to its local telephone exchange by a telephone line. The telephone exchange is able to switch circuits so that any two of the lines on that exchange may be connected together and the users may have a telephone conversation. In addition, local telephone exchanges are interconnected by a nationwide switched network, and by dialling the correct code a user may be connected to a number on any other telephone exchange in the country. An international telephone network also exists that allows international telephone calls to be made.

The speech-frequency electrical signal produced by the telephone transmitter is not in a form that allows it to be directly transmitted over the telephone network and it therefore requires to be processed before it is transmitted.

The following are further examples of information transmission:

1. **News reports**. The block diagram of a news-reporting information system is shown by Fig. 1.3. Textual information, which is inputted by a reporter into a word processor via a keyboard, is processed to remove all errors and to arrange the news item into the wanted format and style. Each finished news item is then transmitted over a communication link to the distant editorial office. Here the information may be further edited and combined with other items obtained from other sources before the final version is passed to either an autocue for a television newsreader, or a printing press for a newspaper.

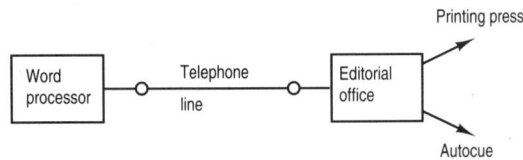

Fig. 1.3 News-reporting system.

2. **Weather forecast**. The block diagram of a weather-forecasting information system is shown in Fig. 1.4. Information about the weather — air pressure and temperature, and wind velocity and direction, at various points around the Earth — is gathered by a number of weather satellites in orbit above the Earth's surface. This information is transmitted to a weather centre on the land where a large computer uses the collected data to produce charts that show what is happening. This information is then interpreted

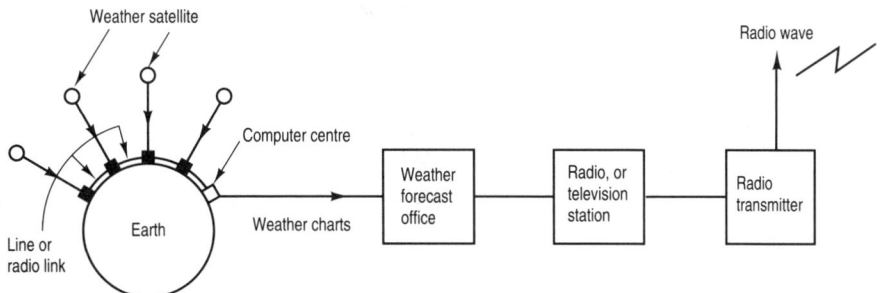

Fig. 1.4 Weather-forecasting system.

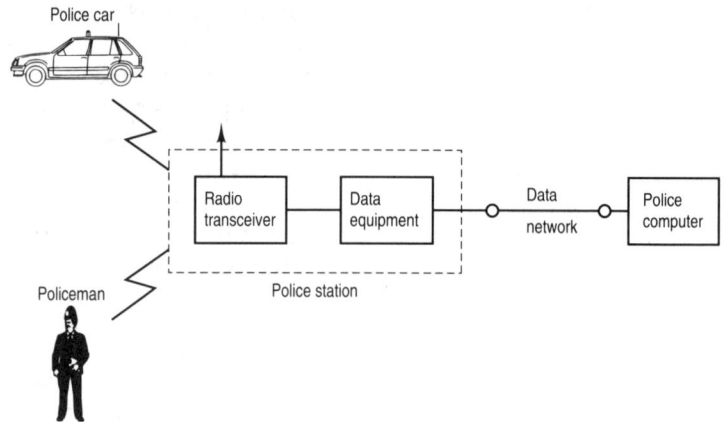

Fig. 1.5 Police computer network.

by weather forecasters to produce forecasts for the various regions of the country. For a television broadcast the weather forecast is given largely in graphical form, but for a radio or newspaper forecast a mixture of text and figures is employed. At the appointed times the television weather forecaster comes 'on-air' and delivers the forecast. The spoken and graphical information is transmitted through the atmosphere by radio waves, is picked up by an aerial and, after processing in the television receiver, is displayed on the screen.

3. **Police computer network**. Every policeman in the country has immediate access to the police computer that is used to store information on known criminals, stolen cars, missing persons, etc. A policeman requiring information can radio to his headquarters and request specific information. The police headquarters has access to the police computer via a private data network and is able to quickly access the computer and obtain the required information (if it is held, of course). Figure 1.5 shows the basic arrangement of the police computer network.

The electronic office

A variety of electronic machines are used in a modern office. These include machines for producing written material, for the storage of information, for the copying of written and graphical information and for the calculation of numbers.

Typewriters

A typewriter is employed to produce typed information, such as letters and invoices. The simplest kind of typewriter is the manual type which is purely mechanical in its operation, although it may have an electric motor, and has no memory. When the typist presses a key on the keyboard a hammer immediately strikes an inked ribbon to make the appropriate mark on the paper. The manual typewriter is cheap, easy to use, and it is perfectly adequate for low-volume general-purpose use.

An electronic typewriter has a limited memory and is able to produce good results with the minimum effort. The print speed is usually either 10 or 12 characters per second. Usually, facilities such as centring, underlining, right-hand justification, and bold print are offered. The characters are not printed immediately a key is pressed: the characters first appear on a small screen and, if necessary, may be corrected before they are printed. The amount of text that may be corrected varies with the typewriter but one-line correction is common, with three-line correction being the usual maximum. Some of the more expensive electronic typewriters are provided with typical word processor facilities such as justification, superscript, and subscript, and may also have a spell-check dictionary. Some models have a small amount of memory ranging from, say, three pages to twelve pages of A4 text. Further details on typewriters are given in Chapter 5.

Word processors

A word processor may be a purpose-built equipment that has no other function, known as a dedicated word processor, or it may be a microcomputer that runs word-processing software. In either case a fairly large memory is provided. A word processor is able to carry out all the functions of a typewriter, electronic or manual, and also offers several other useful facilities such as copy, move, insert and delete single characters or complete lines or blocks of text. Letters and other documents can be created and edited on-screen; they can then be printed out when the operator is satisfied with the displayed text, thus avoiding the need to re-type whole pages of text. Some word processors are WYSIWYG, i.e. 'what you see is what you get', which means that the display on the screen will be the same as the print-out. This feature enables the operator to see the appearance of the finished text.

A document may be stored on a disk for subsequent retrieval whenever it is required and it may then be further edited. A variety of different kinds of printer may be employed in conjunction with a word processor and the selection of a particular kind of printer for a specific application is discussed in Chapter 5. Printing is always bidirectional and a spell-check dictionary is usually provided. In most offices the use of word processors instead of typewriters will considerably increase the productivity of the office. This is because: alterations to a text may be made on-screen avoiding the need to make paper corrections with Tippex or the re-typing of whole pages of text; information that is repeatedly used — such as standard phrases and names and addresses — can be stored on a disk and automatically inserted into a document.

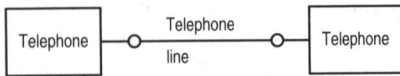

Fig. 1.6 Basic telephone circuit.

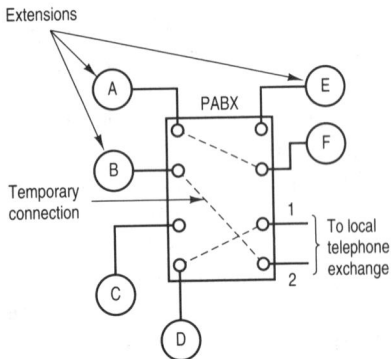

Fig. 1.7 Use of a PABX.

Telephones

The telephone is used in an office to allow speech information to be passed from one person to another. The two parties to a conversation may be situated in the same building or they may be many miles apart — perhaps even in different countries. The basic telephone connection, shown by Fig. 1.6, can be seen to consist merely of two telephones connected together by a two-wire line. Each telephone, known as an **extension**, has a transmitter, a receiver, a calling device such as a bell or a ringer, and a means of 'dialling' a wanted number; the last may be either an actual dial or, for modern telephones, a keypad. There is usually a need for each extension in an office to be able to be connected to any other extension in the office, and to give each extension access to an exchange line. The necessary switching is carried out by an 'in-house' automatic telephone exchange. The basic principle of a **private automatic branch exchange** (PABX) is illustrated by Fig. 1.7. Each extension is connected to the PABX by a two-wire line and any extension can be connected to any other extension by 'dialling' the number of the wanted telephone (the term 'dialling' is still used even for telephones that have a keypad instead of a dial). An extension may gain access to an outside line by 'dialling' a specific number, very often 9; a dialling tone is then obtained from the PSTN and the wanted national telephone number may then be 'dialled'. Modern telephones can now offer a wide range of facilities (pp. 22, 204).

A PABX can be set up with a particular extension designated to act as the 'switchboard'; the operator of this extension answers all incoming calls and switches them to the required extension, or the system may be arranged so that any extension may answer an incoming call. It is easy to change over from one method of operation to the other. Examples of PABXs offered by British Telecom (BT) include: 1 + 4, which has one exchange line and up to four extensions; 2 + 8, which has two exchange lines and eight extensions; 6 + 16, where 6 exchange lines and 16 extensions are available.

Answering machines

Often associated with a telephone is a telephone-answering machine. If an incoming call is not answered within a certain time the machine automatically switches on, passes a message to the caller and then records any response that the caller might make. The answering machine may later be switched into its 'play-back' mode when all recorded messages can be heard. Facilities that may be provided by an answering machine include:

1. **Call monitor**. This allows the user to listen while the machine answers the call and, when the caller has been identified, to decide whether or not to answer the call.

2. **Two-way conversation recording**. This facility allows the user to record any telephone conversation.
3. **Remote message retrieval**. The user is able to check messages held in the answering machine from any other telephone. A voice or electronic tone code is used to instruct the machine to pass any messages to the user.
4. **Memos**. The memo facility allows a message to be left on the machine for another person to hear at some later time.
5. **Answer only**. The machine is programmed to play its message without asking the caller to leave a response.
6. **Call count**. This feature displays the number of messages that have been recorded.
7. **Day/time record**. This facility automatically records the time and date of each recorded message left on the machine.

Facsimile telegraphy

Facsimile telegraphy, or 'fax', is the transmission of photographs, diagrams, documents, etc., from one location to another over an ordinary telephone line. Most fax machines operate over the PSTN and merely plug into a normal telephone socket but they must be provided with a mains electricity supply. A user first dials the telephone number of the destination fax machine and, when a connection has been established, transmits the information. The document to be transmitted is placed into the machine where it is held on the surface of a revolving drum. As the drum rotates the document is scanned by a photoelectric cell and a coded digital electrical signal is produced. This digital signal represents the image to be transmitted. The coded signal is then applied to a modulator which converts the signal into analogue form to enable it to pass over the telephone network. At the receiving terminal the incoming electrical signals are demodulated to convert them back into coded digital form, and the digital signals are then applied to a decoder. The output of this circuit causes the machine to produce a replica of the original document.

Any document that can be photo-copied can also be faxed. A fax machine is able to transmit an A4 sheet of information over a normal telephone line in 20 seconds or less. A fax switch allows a telephone and a fax machine to share a common telephone line. The switch monitors each incoming call and diverts it to either the telephone or the fax machine. Some fax machines are able to store frequently called telephone numbers in their memory; these numbers can then be called by the touch of a few keys on a keypad.

Some fax machines incorporate an answering machine that can be connected when the office is unattended. All incoming calls will then be answered automatically by the answering machine and if any fax message is recognized it will be diverted to the fax machine.

Reprographic equipment

The modern office is usually equipped with systems that can reproduce letters, documents, and diagrams. A **photo-copier** enables a copy to be taken of any written or printed material (within the machine's dimensions) that is placed into the machine. The copier can be set to produce any given number of copies and the lightness/darkness of the copies can be set to suit requirements. Two basic types of photo-copier exist.

In the first type, special paper with a photo-conductive coating (usually zinc oxide) must be used. The paper is negatively charged so that it will attract small particles. The small electric charges will disappear from any points on the paper that are exposed to light. The document to be copied is placed in the machine and is then illuminated by a bright light. The light is able to pass through the document wherever the paper is blank but it is unable to pass wherever the paper has a character, or some other mark, inked on it. A charged image of the document is then set up on the copier paper. A dry powdered ink, known as **toner**, is blown over the photo-copy paper and is attracted to each part of the paper that is negatively charged. The inked image must then be 'fixed', which is done by heating the paper so that the ink melts into the paper to produce a permanent copy.

The second kind of copier works in a similar way but permits the use of ordinary paper. The paper is pressed against a photo-conductive coated material that is an electrical insulator until it is illuminated by light. The material is charged electrically and the original is then illuminated. The material then becomes an electrical conductor at all points that have been reached by light; the charge then leaks away except at the dark areas, which are the inked areas of the original. As with the other method a charged image of the original is thus obtained and fixed in the same way.

Many of the latest types of photo-copier are controlled by a microprocessor so that the equipment is a form of electronic system.

Calculators

Calculators are commonly employed in offices — and nearly everywhere else! — for the rapid and accurate calculations which they make possible. Calculators can be grouped into one of the following five categories:

1. **General**. These calculators are used for such purposes as working out bills and calculating change. They are provided with arithmetic functions such as addition, subtraction, division, and multiplication. A small memory is usually provided.
2. **Scientific**. In addition to the above functions a number of scientific/engineering functions, such as sin, cos, tan, square

roots, powers, etc., are also provided and the memory is much larger. Some of the more expensive scientific calculators are programmable, allowing them to perform even more complex calculations. Some programmable calculators also provide a graphical representation of the result of a calculation.

3. **Financial**. A financial calculator is one that has been designed to be of assistance in financial calculations, such as compound interest.

4. **Desktop and printing**. These kinds of calculators are designed to rest on a desk top and usually have an angled display for easier reading. They may include an electronic diary that is able to store names, addresses and telephone numbers. The printing types can always be operated in a non-printing mode to conserve paper.

5. **Electronic organizers**. An electronic organizer is more than just a calculator; it is more a form of small data management system that is able to store a large number of items of information, ranging from telephone numbers and addresses to financial records and schedules. An electronic organizer has a QWERTY keyboard and is best regarded as a combination of a calculator, an address book, and a diary.

Electronic mail

Electronic mail is a system that provides its users with a cheap and rapid method of passing messages from one point to another. In the United Kingdom a country-wide electronic mail service is provided by British Telecom and is known as Telecom Gold. An electronic mail system may also be used to provide communication within an office using a **local area network** (LAN).

Electronic mail is the telecommunications equivalent of the postal service that delivers letters to the home and to the office. A message sent by post is one that does not need to be delivered within a very short time of its despatch since delivery will take a least one day, and perhaps several days. Users of the electronic mail service send their messages into the system and leave the system to deliver the messages to the destination addresses; this is analogous to posting a letter but delivery usually takes only a few minutes. When the recipients check their electronic mailboxes they will find the messages waiting there for their attention.

Electronic mail has some advantages over the telephone:

1. A message may be sent when it is convenient to the sender; it can take quite a long time to get through to a telephone and even when the called number is reached the person who is wanted may not be in his office. The difficulty is that for a telephone conversation to take place both parties to the conversation must be present at their telephones at the same time. This is becoming less of a problem since so many people now have the use of a mobile telephone but the difficulty does not exist at all if electronic mail is employed.

2. Orders, invoices, and receipts are often sent from one office to another or between people in the same office. Such documents are often prepared using a computer and are dealt with later by the recipient using another computer. Both time and money can be saved if the information is transferred directly from one computer to another using electronic mail.

3. Arranging business meetings is made easier if people keep their diaries on the system. The software can find the most convenient time for all concerned and notify each person of the selected date and time.

4. Information sharing becomes possible and within an office telephone calls can be answered and queries dealt with by any one of several persons, all of whom have immediate access to the same information.

5. Broadcasting allows a message to be sent to everyone connected who has access to the system.

Bulletin boards

A **bulletin board** is the electronic equivalent of a noticeboard or a pigeonhole. Any user is able to access a central computer from a personal computer, or computer terminal, and leave a message for one, or more, other user(s). Files of all kinds can be left on the system and exchanged between users. Both messages and files are stored in the central computer's hard disk until they have been read by the recipient, after which they may be deleted. Recipients must access the central computer before any message can be read and may copy any files back to their own hard disk. Some bulletin boards are open and anyone who has the appropriate number can access the system and read the messages. Often names and passwords are assigned to restrict access to authorized users. The use of a bulletin board offers one big advantage over the use of the telephone; reading and replying to messages can be done when it is convenient rather than when the telephone rings or when the other party is known to be in his or her office.

Document storage

Offices generate a tremendous amount of information of all kinds and much of this information must be stored for future reference. The information can, of course, be written on paper and then stored in a number of filing cabinets, and this is still the practice in many offices. Paper storage is cheap, portable, and easily copied but it takes up a large amount of storage space, and difficulties may arise with gaining access to a particular piece of information. The problem gets worse with each increase in the amount of information stored and, consequently, the storage of information using electronic and magnetic means is commonly employed. Electronic document storage methods include computer memory systems, microfilm, microfiche, tape streamers, and document and image processing (DIP).

Computer storage

Computers are commonly employed to keep records and to manage large amounts of data in both tabular and graphical form. The records might include details of personnel, stocks of goods held, etc., while the data might include monthly sales figures, staff attendances, salaries, etc. A computerized collection of records is known as a **database**. Software is used to enter data into, or to retrieve data from, a database. Database software allows selected records to be rapidly retrieved, records to be sorted into a specified order, records to be modified and/or added to, and so on. In a database records are kept in a file of forms; each form in a file has both a label to identify it and a field in which the data is held. A **spreadsheet** consists essentially of a number of columns and rows which form cells. Data may be entered in any cell and calculations may be performed on that data. Each column in the table is labelled at the top with headings that correspond with the field labels in a record. Collectively, database and spreadsheet software packages are known as **application software** which is dealt with in more detail in Chapter 4.

The data will be stored in either a **floppy disk** or a **hard disk**. Perhaps the main advantage of computer storage is its ability to retrieve instantly any required piece of information.

Microfilm, microfiche, tape streamers, and DIP

Much of the information used by an office is required only occasionally for reference, or other purposes, and storage can be a very bulky problem if the information is stored on paper. Microfilm and microfiche provide two ways of rapidly storing and retrieving large quantities of data. Microfilm reduces large quantities of paper records into a few rolls of microfilm, and microfiche reduces the records into a box of large transparencies. The block diagram of a microfilm system is shown by Fig. 1.8. The data to be stored is converted into characters which are displayed on the screen of the monitor one page at a time. A microfilm camera is then used to photograph each page of the display using 16 mm film. Hard copy data can also be filmed. The filming is carried out at a very high speed, typically some 500 000

Fig. 1.8 Microfilm system.

characters per second. When the film has been developed it is stored in a microfilm library until the data is wanted. The data can be accessed by placing the film in a microfilm reader. This is a computer-controlled viewing device that includes an index of each film, giving both its contents and its location in the library. The reader enables any page of data to be converted into a video image and be viewed on a screen by the user keying in its reference number. A hard copy of any data can also be made if required. If necessary, the displayed information can be modified and then re-recorded and stored for future reference.

Other storage systems employ **microfiche**. Microfiche is a sheet of film, of typical dimensions 10×7.5 cm, on which a large number of pages, perhaps as many as 80, of data can be recorded. A fiche then contains an array of data pages, each of which is a miniature of the original document. This method of data storage is best suited to storing documents held on paper and computer-generated graphics and text. The data stored on microfiche can also be retrieved by reading it with equipment similar to that used with microfilm.

A number of advantages are claimed for the use of these techniques:

1. The speed of outputting the data is much higher than any printer could manage.
2. As well as the usual textual material, diagrams and charts can be reproduced.
3. Film storage is relatively cheap.
4. There is no problem if a printed copy of the data is wanted.
5. The physical dimensions of microfilm and microfiche are small and hence large quantities of data may be stored in a small area.

There is one obvious disadvantage inherent in the use of microfilm or microfiche; this is the need for special viewing equipment to allow the stored data to be accessed. This equipment is rather expensive and so the system is only employed by organizations that have large amounts of infrequently used data to store, such as an insurance company storing details of policies or a newspaper storing a copy of each edition. Microfiche is particularly employed when information is to be sent from a central storage point to a number of locations where the data is wanted at regular intervals of time. The data on microfiche can be sent by post to its destination; a common example of this is information sent to garages by car manufacturers, or by spare-part manufacturers.

Tape streamers

A **tape streamer** is a cassette storage system that is used to provide back-up storage of the data held on a computer hard disk (see p. 56). It is only useful for storing groups of files together since it takes some time to copy material to and from the tape.

Document and image processing

Document and image processing (DIP) is a method of storing documents. A document is entered into the system by scanning it with an image scanner. The image scanner works in a similar manner to a photo-copier and copies one page of information at a time. Documents can also be directly entered into the system from a fax machine. As a document is entered it is processed by a personal computer and reduced in size so that it can be stored on an optical disk. At the same time an enlarged version of the data is displayed on the screen. A 5.25 inch optical disk is able to store up to 800 million bytes of data; this is the equivalent of the contents of a four-drawer filing cabinet.

Commerce

There are a number of applications of information technology in the field of commerce with which everyone will be familiar. These include bar codes on goods in shops, credit-card terminals, and debit-card terminals.

Bar codes

Bar codes are a key element of modern information technology and they are found in the kitchen, in shops on the sides of packets and tins of food, in libraries and bookshops on the covers of books, in hospitals on packets and bottles of drugs and medicines, and in many other places. To the uninitiated they appear to be a number of strange black and white markings, but in fact each code is unique and identifies the product on whose wrapping or tin it is printed. Before bar codes were employed every item on the shelves of a supermarket had to be individually price marked with a sticker. Now that all products have a bar code printed on their package this marking is no longer necessary. A bar code does not give the price of an item (which is a variable quantity — usually upwards!), but it does identify the product. The actual price of the item is indicated on the shelf and is held in a computer database that is linked to each checkout. When a purchase is scanned its bar-code number is read and supplied to the computer. The computer recognizes the bought item, looks up a price list held in memory, and then causes a description of the item and its price to be printed out on the shopper's receipt. A running total of the money owing is also printed.

A bar-code symbol is a pattern of black bars of varying widths and spacings. The pattern can be read optically by passing a small light beam (a laser beam in supermarket scanners) over the bar-code pattern, when all the light is absorbed by the black bars but some light is reflected by each white space. The bar-code scanner senses the differences in reflected light and converts these differences into digital electrical signals which are then transmitted to a computer. In some systems a black bar indicates logic 1 and a white space indicates logic 0; in other systems a wide bar or space indicates 1 and a narrow bar or space indicates 0. The former method is known as **contrast encoding** and the latter as **bar-width encoding**. Industrial applications generally employ bar-width encoding and commercial applications employ contrast encoding.

Besides keeping a check on the goods sold the computer also transmits details of each transaction to a main computer that keeps a note of stock levels. A typical bar-code label can be seen on the back cover of this book.

Bar-code labels are increasingly employed today because they offer the following advantages:

1. **Speed**. It takes a hand-held bar-code scanner about 2 seconds to scan a bar code and this is at least three times faster than a skilled operator could enter the same data manually.
2. **Accuracy**. On average it has been found that manually entered data has one error for every 200 entries, whereas bar-code entry has a typical error rate of one in 300 million.
3. **Ease of use**. It is very easy to operate a bar scanner, as anyone who has been in a supermarket will know. It takes only a few minutes to train a check-out operator to use the system.

These advantages lead to greater efficiency at the check-out giving a greater throughput, lower labour costs, and shorter queues.

For a bar-code system to work accurately and efficiently the labels used must be produced in a way that reduces potential problems caused by dirt, abrasions, smudging, smearing, labels falling off, and so on, and most labels are produced using a computer-controlled photographic process.

The design of a bar code involves a consideration of the following factors: (a) the bar-code language to be employed, (b) the width and height of the narrow and wide bars, (c) the number of characters per inch, and (d) the bar orientation, i.e. horizontal, vertical, or both on the same label. The bar-code readers fitted in the supermarket check-outs must be able to read any of the bar codes that may appear on the goods stocked and hence it is necessary that some standardization of codes, dimensions, etc., for the bar-coded labels is agreed upon by the various manufacturers. In the UK, food and most other items use a bar code that starts with 50, but book bar codes start with 978 and magazine bar codes with 977.

Point-of-sale terminal

A **point-of-sale terminal** (PoST) is able to identify bar-coded goods automatically and produce a bill for the customer using price information stored in the system. The terminal can also transmit details of each sale to the stock control computer where a note is kept of the stocks held of each item sold, and when the stock is low the ordering of replacement goods can be carried out automatically. Each terminal can work alone having its own program and data storage facilities, or one terminal can be operated as a master which supplies all other terminals with programs and data. Data will be exchanged with the company's main computer at regular intervals.

Some terminals also incorporate **electronic fund transfer at point of sale** (EFTPoS) facilities.

Credit card terminals

When a customer offers a credit card, such as Mastercard or Visa, to pay for goods purchased the shop assistant 'wipes' the card through a counter-mounted card reader. The reader reads the magnetically stored information on the card, e.g. its number and the card-holder's name, and prints out a voucher which the purchaser is required to sign to verify the signature on the card. A copy of this voucher then becomes the purchaser's receipt. The credit card terminal is connected to the credit card company's computer centre by a telephone line and details of the transaction are sent to this centre. Here the details are entered into the customer's account and added to the bill which is sent to the customer each month.

The newer debit cards, such as Switch and Delta, operate in a similar way except that the amount spent on a purchase is automatically debited from the customer's bank account within a working day or two. Increasingly, stores are allowing customers to obtain cash from them using a debit card.

The use of electronic fund transfer at point of sale (EFTPoS) reduces: the amount of money that customers need to carry around with them with the ever-present risks of theft or loss; the amount of paperwork that banks and building societies have to handle when goods are paid for by cheque; the amount of paperwork a retailer has to handle when a credit card is offered; the time that elapses before the money deriving from a transaction is credited to the retailer's bank account.

The block diagram of an EFTPoS system is shown by Fig. 1.9.

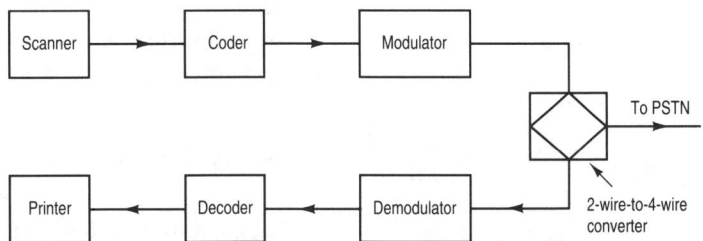

Fig. 1.9 EFTPoS terminal.

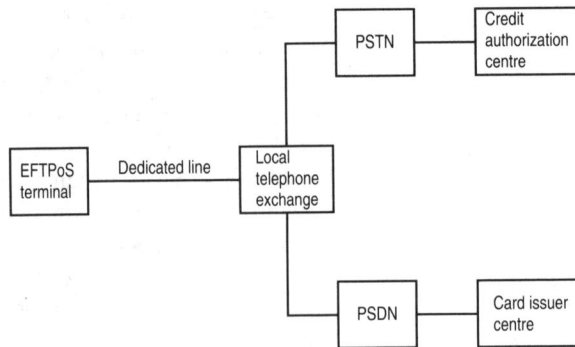

Fig. 1.10 On-line EFTPoS system.

Clearly, it would not be economic for each retailer to have a link to each company that issues a credit/debit card and a national EFTPoS UK system has been set up. An EFTPoS terminal may be either 'on-line' or 'off-line'. On-line operation gives instant communication with the system's computer so that there is no need for data to be stored at the terminal and also any stolen cards can be immediately identified. Off-line operation has the advantage of storing data until night time when telecommunication charges are lower. The basic arrangement of an on-line system is shown in Fig. 1.10. Each terminal is connected by a **dedicated**, i.e. permanent private, line to the local telephone exchange. From here a connection is set up via the PSDN to the card issuer's computer. If credit authorization is necessary the credit authorization office is accessed by a speech connection via the PSTN.

Finance

Every transaction of money is a form of information transmission and a record must be kept of each one. The finance world of banks, building societies, insurance companies, and the Stock Exchange make great use of information technology. These institutions have installed extensive private data networks to transfer data rapidly from one point to another, to calculate wages and salaries, to deduct tax, to buy and sell shares, and so on. Cheques have the cheque number, the branch sort code, and the account number printed in special lettering so that they can be read by an optical scanner and the data thus obtained fed into a computer. The actual amount of money for which a cheque has been written must be keyed in by an operator. The computer keeps a record of each bank account and automatically sends a monthly statement to each account holder. Electronic fund transfer is used to move money from one bank or building society to another and wall cash dispensers provide a quick and easy way by which people can draw money from their account.

Electronic fund transfer

In the UK the transfer of money between banks and building societies

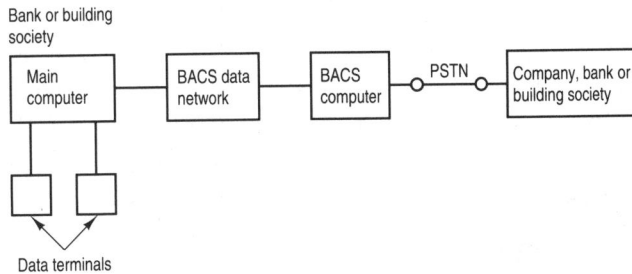

Fig. 1.11 BACS system.

is handled by a system known as the Banker's Automated Clearing Service (BACS). The BACS system is also used by both the Government and many companies to pay the salaries and pensions of their staff and former staff into their bank or building society accounts. The system is also used by some companies for the payment of accounts. Figure 1.11 shows the basic arrangement of the BACS system. The advantages claimed for BACS include: greater efficiency; improved cash flow because of same-day debiting and crediting; smaller bank charges; payments are not dependent on the postal system; and transferred funds are available for use on the due date without the delay involved in waiting for a cheque to clear.

Automated telling machine networks

An automated telling machine (ATM) network connects a large number of cash-dispensing machines to a data network and thence to a large computer. In the UK three large ATM networks are commonly seen in the high street: (i) the LINK system that serves several building societies, including the 'big three': Abbey National, Halifax, and Nationwide; (ii) a network that serves Barclays bank, Lloyds bank, the Royal Bank of Scotland, and the Bank of Scotland; (iii) a network that serves the Midland and NatWest banks, the TSB, as well as several smaller banks. The two bank networks also allow Mastercard and VISA cards to be used to obtain cash. Each cashcard has a magnetic strip that carries a unique code to identify the account of the holder to the network. The card user must also key in his **personal identification number** (PIN). If the two numbers tally a request for cash at a bank's cash dispenser will be transmitted to that bank's computer. The basic arrangement of an ATM system is illustrated by Fig. 1.12. If the request has been made by a customer of that particular bank the computer will check the account of the card user and if there are sufficient funds it will send back a signal authorizing the ATM to pay out the cash. If the request has been made by a customer of another bank or of a building society connected to that ATM system the request is passed on to the other organization's computer for authorization. A cashcard can also be used to check a bank balance and to order a statement.

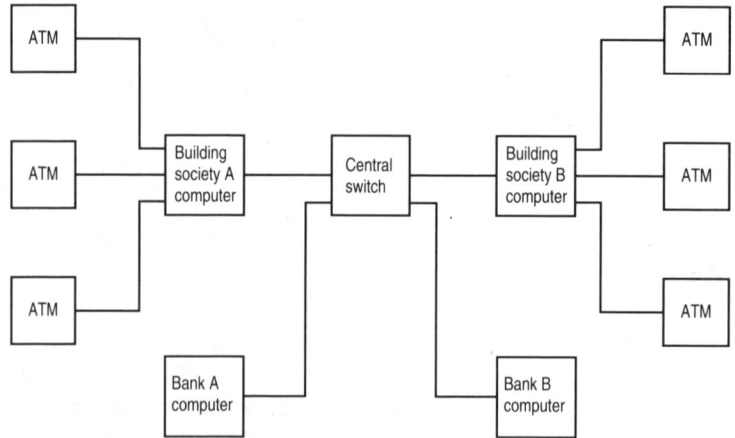

Fig. 1.12 Cash dispenser network.

Electronic data interchange

Increasing use is being made of **electronic data interchange** (EDI) by business. EDI is the electronic transfer of purchase orders, delivery notes, invoices, receipts, etc., between businesses. The use of EDI considerably reduces the amount of paperwork that needs to be done in an office and gives an increase in the speed with which documents are prepared and processed. It also avoids the delays inherent with the postal service.

Smart cards

Smart cards have begun to appear in a number of applications such as banking and telecommunications. A smart card looks just like an ordinary plastic card but within it is a silicon chip that contains a program which is able to store and calculate details of all the purchases for which it is used. When used the smart card is inserted into a card reader, and contacts on the card connect with circuitry in the reader. All current smart cards operate as slaves; their internal program acts only to interpret commands received from the reader. The producer of a smart card makes a deposit of resources in the card, such as an amount of money, a number of telephone calls, or an amount of travel on a public transport system. The card user can draw on these resources as, when, and where required. Periodically details of the resources used are read from the card by a machine in the user's bank and are then fed to a computer which debits the correct amount from the user's bank account.

Point-of-information systems

A **point-of-information system** is used to provide information to visitors to an establishment. The user touches a point on the screen of a monitor to select his choice from a displayed menu. The required information is then displayed on the screen. These systems are to be seen in such diverse places as museums, picture galleries, civic centres, and town halls. They are used to give visitors guidance to enable them to find a particular department, and to give information on the work done in the building or on whatever the visitors are about to see.

Publishing and printing

The normal path of a magazine, book or other printed material is author, editing, printing, warehousing, marketing, and (hopefully) selling. All of these steps can be assisted with the aid of information technology. The task of the author is considerably eased by the use of a word processor and the finished script can be delivered to the publisher on a computer disk as well as the hard copy (on paper). Typesetting firms use computerized equipment that allows text produced on one particular kind of word processor to be converted to the form required by the computer-controlled typesetting equipment. Editorial instructions marked on the hard copy can be carried out on-screen and the manuscript amended to remove all (noticed!) errors, and set the required type sizes and text and figure layouts. Figures are usually drawn separately and inserted into the final page layouts since the amount of computer memory needed to handle graphics is very large. The camera ready material is then sent to the printers for the production of the book.

Printed magazines are probably sent to wholesalers for distribution to shops since most magazines are published either weekly or monthly and old copies are of little value. Books are stored in a warehouse and copies are sent to bookshops when ordered. A computer is employed at the warehouse to keep track of the number of copies of each book left in stock as the books are sold. The computer can be programmed to produce a monthly print-out of the number of copies of each book sold, and to inform the publisher when the stock level has fallen to a number at which, perhaps, a re-print should be set in motion.

Desktop publishing (DTP) software has been developed to enable personal computer users to produce publications of a quality that is comparable with that obtained from a commercial typesetter. It is used to produce brochures, news-sheets, and magazines. A desktop publishing package allows a variety of typefaces to be used, it can incorporate any graphics and pictures that are wanted, and gives the user control over the final appearance of each page. Usually a **laser** printer is used with a DTP package in order to get a high-quality result.

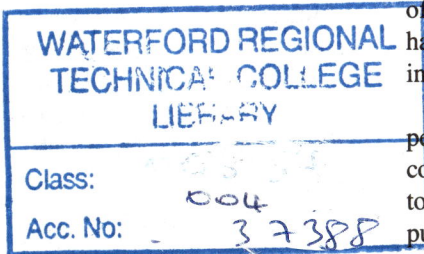

Communications

Information technology includes the transmission of information from one location to another, and this means that telecommunication systems are an important aspect of IT. The telecommunication systems employed in IT include the public telephone service, radio and television broadcasting, mobile and cellular radio, data communication, and facsimile telegraphy (fax).

Telephony

The telephone service, provided in the UK mainly by British Telecom, is used for the transmission of both speech and data information. The public switched telephone network (PSTN) of the UK has been modernized in recent years and now relies heavily on digital techniques and computer control. This is discussed in Chapter 10. The modern telephone includes a silicon chip, or IC, which allows the user to obtain a required number by pressing the appropriate keys on a keypad instead of rotating a dial as with older telephones (although the process is still referred to as 'dialling a number'). The ICs also provide some useful facilities such as auto-'dialling' and last number recall. Last number recall means that the telephone automatically remembers the last number 'dialled' and if a key is pressed will re-'dial' that number. A selection of telephone numbers can be stored in a telephone: a number is first keyed-in so that the instrument captures the number, and a 'store' key is then pressed to store the number. When, some time later, this number is to be 'dialled' again the 'memory' key is pressed to retrieve the number. The telephone then processes the stored information by converting it into the form of the calling signal required by the local telephone exchange, either a number of on—off direct current pulses, or pairs of pulses at two different frequencies.

Other facilities that a modern telephone may offer include: **display**, by which the user can see the number called in a display panel; **secrecy** or **mute button**, which allows the user to speak to someone else in the room without the caller hearing the conversation; **on-hook 'dialling'**, which allows the user to dial a number without lifting the handset; **call screening**, which allows the user to listen ot a call before deciding whether or not to answer the call; **memo message recording**, which allows the user to record a message for others to hear if he or she wishes to leave the office; **two-way recording**, which allows the user to record complete telephone conversations as they take place.

Cellular radio

Cellular radio is a computer-controlled mobile radio system that allows users to communicate with each other and with customers of the PSTN as they travel around the country by motor vehicle (p. 226).

The country has been divided up into a large number of **cells**. In the (rough) centre of each cell there is sited a low-power base station. Each base station is allocated a number of telephone channels over which it can transmit to, or receive from, any **cellphone** in that cell.

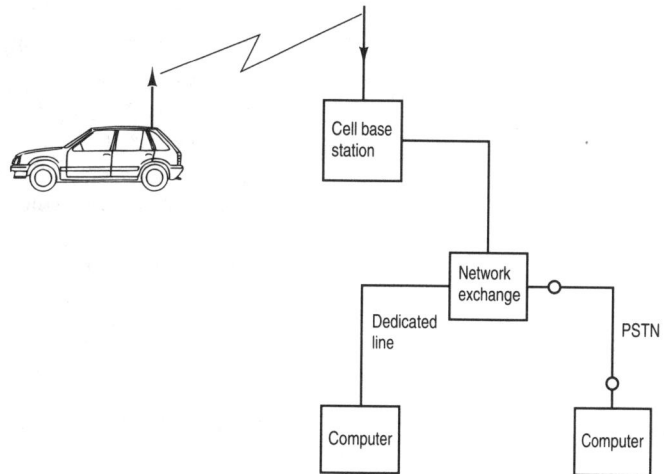

Fig. 1.13 Cellular radio.

The geographic size of each cell varies in diameter from about 2 km in city areas to about 30 km in rural areas. All the base stations are interconnected to one another and to the PSTN by land lines, as shown by Fig. 1.13. The cellular network is often known as **cellnet**.

When a user makes a telephone call using a cellphone he or she is allocated transmitting and receiving frequencies by the base station of the cell in which he or she happens to be at that moment. The caller then keys the wanted telephone number and this is transmitted to the base station over a control channel. The base station obtains the wanted number on the PSTN and connects it to the calling cellphone. If the car is driven from one cell to another while a call is in progress the original base station automatically switches the call to a new base station and this station allocates new frequencies to the connection. The switching takes place so rapidly that the parties to the telephone conversation remain unaware that anything has happened. When a PSTN customer calls a cellphone the control computer pages all cellphones over a control channel. When the wanted cellphone is found, cell frequencies are allocated and the call is set up by the appropriate base station.

Cellular radio is at present an analogue system designed primarily for speech communication. A new digital data system is being introduced that has the ability to input, display and print out alphanumeric data at both ends of a link between a mobile and a control centre.

Telex

Telex is a method of transmitting textual information between two telex machines and has been in operation for many years. Telex uses an independent switched network and is unable to transmit lower-case characters, drawings or graphs. Since its speed of operation is low, only 300 bits per second, it has been partly replaced by an improved system known as **Teletex** which is able to operate at speeds up to 2400 bits per second with both upper- and lower-case letters. A variety of equipments, such as electronic typewriters and computers, can act as Teletex terminals. Teletex operates over the PSTN. Telex allows textual messages to be sent or received to or from most countries in the world. Telex is very reliable because of agreed world-wide standards of compatibility and the fact that both national and international exclusively Telex networks are employed. Telex gives the user written proof that a message has been received by the correct recipient, and in many countries a Telex is recognized as a legal document. The **Telex Plus** service gives repeated delivery attempts and will broadcast a Telex to up to 1000 different destinations. However, both systems are being replaced by fax.

On-line databases

A **database** is a collection of data that can be accessed and used by a number of different users. An on-line database is one that customers of the service can access via either a PSTN or a PSDN connection, or via a private data circuit, to obtain information. An example of the former is British Telecom's Prestel service. prestel is a **viewdata** or **videotex** system in which information is displayed on-screen in a large number of **frames**. An index frame, similar to those used with Ceefax and Oracle, shows the information that is available and the number that must be keyed in to gain access to particular information.

Prestel services include: **Phone Base**, which is a database of all UK business, residential and government telephone numbers; **Electronic Yellow Pages**, which is a database that lists UK businesses; **Cityservice**, which is a database of news, City prices, and economic and financial reports; and **Infocheck**, which is a credit database that contains basic information on every UK company. Also, **Mailbox** is Prestel's version of electronic mail.

The basic block diagram of the Prestel system is shown in Fig. 1.14. The user has a telephone to which a television receiver or a monitor is connected via a suitable interface. This interface must include a keyboard, a **modem** which converts data signals from digital to analogue, or vice versa, an auto-'dialler', and circuitry that is able to convert received data into a form that the TV receiver can display.

Private databases include the police national computer, and the Driver and Vehicle Licensing Centre at Swansea, which holds details of all cars and their registered owners in the country.

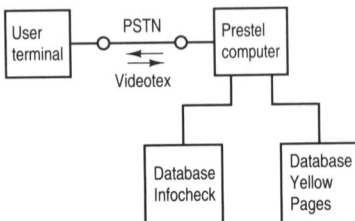

Fig. 1.14 Prestel system.

Broadcasting

Radio and television broadcasting must be the best-known examples of information technology since every home has either a radio receiver or a television receiver, and more probably both. Programmes are broadcast to entertain and inform by the BBC and by the various commercial television companies. In addition, teletext systems known as *Ceefax* and *Teletext* are broadcast along with the television signals; these are menu-driven database systems that offer the user free access to information of many kinds. The user employs a remote control unit to request the page of information required. The two systems are not interactive, i.e. there is no communication from the user to the computer.

Data communication

There is frequently a need for two computers to communicate with one another, or for a computer to communicate with a data terminal. The two computers might be located within the same building or they might be many miles apart. The transmission of data between two computers, or between a computer and some kind of data terminal, is known as **data communication**. Today most of the information required by commerce and industry is stored in computer memory systems. If this information is to be accessed it is much quicker, and more convenient, to transfer the information electronically over a communication circuit than to put the information on to paper and send it by post or by hand. The pricing mechanism of tele-communication systems allows large amounts of data to be sent at relatively low cost, particularly if a high-speed system is employed.

Data circuits allow data held in a distant computer system to be accessed locally. The data circuit may be set up temporarily over either the PSTN or the PSDN, or it may use a private leased line between the two points, or it may be a connection set up via a private data network. A dedicated line, or network, is one that is exclusively used for one purpose. A computer that has been dialled-up over the PSTN is said to be 'on-line' and a database so accessed is said to be an 'on-line database'. Electronic mail is a form of data communication. Data communication is the subject of Chapter 11.

Industry

The stages in the production of a manufactured product, large or small, are as follows:

1. *Design the product.* Market research will normally have been carried out to determine whether there is a demand for the product and the design may be based upon the data obtained from the

research. The actual design may be carried out, wholly or partly, with the aid of a computer with the appropriate software. This is known as **computer-aided design** (CAD). The use of CAD gives (a) increased productivity, because designs can be checked faster and more easily and documentation is easier to produce, and (b) the ability to produce more complex designs.

2. **Produce the product**. Production will be based upon the anticipated number of orders for the product, bearing in mind the need to maintain an adequate stock to meet demand. The planning of the production process is done in two main parts: (i) **scheduling** is the planning of the times when different steps in the manufacturing process will be carried out and ensuring that necessary material and parts are available; (ii) **loading** means deciding which machines and/or workshops should carry out these various tasks. This planning is often carried out with the aid of a computer.

3. **The actual manufacture**. This comprises the making of parts from raw materials and their assembly, together with any bought-in parts, to make the product. Much of the manufacture in a modern factory is carried out automatically by computer-controlled machines. This automation of the manufacturing process is known as **computer-aided manufacture** (CAM).

4. **The manufacturer's products**. These are stored in a warehouse until sold to a shop.

Overall control of the manufacturing process in a factory may be vested in a **data-processing system**, which may be used to perform a wide variety of tasks, including:

(a) The processing of customer's order both to generate sales documentation and to determine how many items should be manufactured at a given time.

(b) Maintaining a stock record of the total goods-in and goods-out.

(c) Calculating the quantities of raw materials and bought-in parts that must be ordered and producing all the necessary purchase orders.

Computer-aided design

Computer-aided design software packages enable the user to produce engineering drawings and diagrams. The software is used in a very similar way to a word-processing package since the folowing facilities are provided:

1. Items can be inserted, moved, or rotated on screen or copied from the screen.

2. Existing drawings stored on disk may be inserted into new diagrams, avoiding the need for items used in more than one

product to be repeatedly drawn. This reduces both the time taken to produce a drawing and the probability of error.

3. The finished drawings, diagrams, etc., can be printed out in whatever format, colour or size is required.

A CAD package may include the equivalents of such drawing aids as a compass, a protractor, and a ruler, and allows different line widths or graph grids to be rapidly drawn. Any part of an on-screen drawing can be greatly expanded so that it may be examined in more detail and/or operated on with greater accuracy, and textual information, such as physical dimensions and component values, may be typed in at the keyboard and then converted to the appropriate size. To make the drawing of required items easier an input device known as a **mouse** is generally employed. In electronics, for example, CAD is employed for the design of printed circuit boards and for the design of integrated circuits.

Computer-aided manufacture

Computer-aided manufacture is the use of a computer to control the manufacture of a product. A machine that is under the control of a computer is known as either a **computer-numerically-controlled** (CNC) machine or as a **robot**. CAM uses the production plans produced by the data-processing system and the designs produced by the CAD package to determine (a) which parts should be made on which machines, and (b) the timing and the order of the manufacture of those parts.

A CNC machine is able to carry out the machining of parts. The software control of processes such as cutting, drilling, and turning leads to the production of very accurately machined parts. The assembly of the various parts that go to make up a product is carried out by a robot. A robot may be able to handle materials, place them correctly into a CNC machine, remove the finished parts, and, lastly, assemble the parts into the finished product. Since all the parts of a CAM system are under the control of software it is relatively easy to alter the role of any machine in the system allowing variations in the product to be easily accommodated; for example, different coloured cars. It also allows a factory to be operated on what is known as a 'just-in-time' basis; this means that products are made on demand instead of being manufactured continuously and then held in store until sold. This method of working has the advantage that it reduces stock-holding costs.

In the home

Information technology is used in the home for domestic appliances, for entertainment, for personal information handling, for security purposes, and as an aid to learning.

1. **Domestic appliances**. Increasingly the operation of domestic appliances is controlled by microprocessors. At present, this application of IT is mainly used for automatic washing machines, for microwave ovens, and for some of the more expensive cookers.

2. **Entertainment**. Entertainment in the home includes the use of radio and television receivers, the video recorder, the camcorder, the CD player, the cassette player, and games played on a home computer. Dedicated electronic games are also available.

3. **Security**. Increasingly sophisticated home security systems are available that are able to detect a break-in to the home using one of several different technologies. The more complex systems are computer-controlled and can be programmed to guard selected doors and rooms, and to give an alarm and/or telephone the police if an intruder is detected.

4. **Personal information handling**. The personal computer can always be employed to carry out such tasks as maintaining home accounts and personal finances, storing the names and addresses of friends and acquaintances, keeping tax details, and so on. Information on money matters, stock exchange prices, etc., can be gained from the teletext pages broadcast by the BBC and ITV.

5. **Learning aids**. A home or personal computer can always be used as a learning aid. Programmes can be bought that will teach a language or some aspect of mathematics, show how various tasks should be performed, teach keyboard skills, and so on. The Open University broadcasts programmes on BBC2 which are supposed to be viewed in conjunction with printed material supplied by post to students studying for a degree.

2 Digital computers

A digital computer is a machine that is able to input data, process the data in a manner determined by a **program**, and then to output the result(s) of the processing to a peripheral device. A program is a set of instructions which a computer is able to obey. Each instruction performs a relatively simple operation, such as 'move the contents of memory location 1000 to the X register'. Characters are represented by a combination of discrete voltages that indicate the two binary numbers 1 and 0. A voltage that can only have one of two possible voltages is known as a **binary digital voltage**. The computer is only able to understand the program instructions if they are written in **machine code** but since this code consists of long strings of binary digits 1 and 0, known as **bits**, it is difficult for a human to write or to understand. Therefore, the program will be written either in a low-level (assembly) language, or, more probably, in a high-level language. The program must then be converted into the wanted machine code by another program that is held inside the computer; this program may be either a **compiler**, or an **interpreter**, or an **assembler**, depending on which language is used.

A computer may be used to carry out mathematical calculations, to control industrial processes, to maintain a database, to act as a word processor, and for many other applications. A computer may be used to perform a variety of tasks by loading it with different programs. A computer terminal consists of a keyboard by which to input data to the computer, and a monitor, and perhaps a printer, to display the computer's output information. A terminal may be either **intelligent** or **non-intelligent**. An intelligent terminal is provided with some processing power which enables it to carry out some relatively simple computing tasks. On a mainframe computer smaller microcomputers are often used as intelligent terminals.

A computer must be provided with some means of storing data and programs on both a short-term and a long-term basis. Some of the storage, or memory, is provided internally to the computer while the remainder is provided externally. Memory devices are broadly divided into two classes according to the type of access that is provided. A **random access memory** allows any location within the memory to be accessed within the same short time, whereas a **serial access memory** is one in which data can only be accessed in the order in which it has been stored.

A computer may be either a **user-programmable** type or a **stored-program** type.

User-programmable computers

A user-programmable computer may have its **operating system** stored internally on an internal memory device known as **read only memory** (ROM), or the operating system may be stored on a **disk**, which must to be loaded into the computer. [The term **load** means to have a program, or data, transferred from external memory into the computer's internal memory.] The operating system is a complex program that controls the computer's circuitry and peripherals, such as the keyboard and the monitor. The operating system acts on the instructions entered into the computer by an **application program** (Chapter 4) for a particular task to be carried out. This kind of computer is widely used in business, commerce, industry, retailing, and in the home for many varied tasks. These tasks can be categorized as being either in **real-time mode** or in **batch-processing mode**.

Real-time processing

Real-time processing means that the input data is processed so quickly that the results of the computations are outputted rapidly enough to be able to influence the action in progress. Real-time operation is therefore always used in conjunction with control systems. For example, in a computer-controlled factory process the data needed by the computer is continually inputted to the computer. The data is processed and the results outputted and applied to the controlled machinery to make any necessary corrections to the process. Any files containing records, figures, and other data must be held in random access memory and are permanently **on-line**. In a real-time system data may be entered at any time and files are immediately up-dated. Real-time operation requires a large immediate access memory and data links to all terminals.

Real-time computer applications include: the control of industrial processes, booking systems, such as those used by airline and package holiday firms, which when accessed by a booking agent give an immediate indication of vacancies and also immediate confirmation of a booking, often printing out the ticket(s) there and then, the Stock Exchange, where the buying and selling of shares and the up-dating of the two FT indices are computerized, and the control of automatic telling machines (ATM), i.e. the well-known wall cash dispensers. Booking systems are often known as **interactive systems** since the response of the system need not be more or less instantaneous, just rapid by human standards.

Batch processing

With batch processing an amount of data is collected together before it is presented at a convenient time to a computer for processing. The collected data may be inputted from either magnetic tape or from magnetic disk and the results of the computations will be printed out. This means that a program is not run and its data is not processed until it is entered into the computer. Each task is entered in its entirety and stored on a hard disc; it is then held in a queue and run at some later time under the control of the operating system. A particular task may remain in a queue for minutes or even hours, depending upon the amount of data to be processed.

Batch-processing tasks include the following: the storage of records such as technical data, medical histories, and staff names and addresses and work records; inventory control; banking — the transfer of monies from one bank to another, maintaining details of customer's accounts including standing orders, direct debits, and processing cheques; billing and invoicing — bills and invoices can be worked out by the computer and printed with the name and address of the customer. Batch processing is much cheaper than real-time operation in its demands on hardware and software and so it is used whenever possible. A typical example of batch processing occurs in a large department store. Details of each day's sales in each department are stored on disk or tape, and at the end of each day are taken to the computer centre. The data is fed into the computer at a convenient time when it will calculate the day's takings, the existing stock levels of each item sold, re-order any items that have reached a low stock level, and so on.

Batch processing of large amounts of data requires the use of a large computer such as a mainframe. It does not need a high-speed computer because the processing speed will always be faster than the speed with which data can eventually be entered or with which the results can be printed out.

Stored-program computers

A stored-program machine, often known as a **dedicated computer**, will immediately perform a particular task as soon as it is switched on. The necessary program is stored internally and is instantly available. The machine is only used for one specific task, some examples being: a car — to control the dashboard display, to control the engine settings, to check that seat belts are worn, etc.; a camera — to control the focus, exposure, and shutter speed; electronic games.

The relative advantages and disadvantages of the two types of computer are:

1. A stored-program computer does not need to load a program before it can operate since the necessary programs are stored in ROM.
2. A stored-program computer will perform its given task immediately it is switched on, but it is limited in the number of different tasks that it is able to carry out.
3. A user-programmed computer takes some time to load programs before operation can take place.
4. The number of tasks that a programmable computer can undertake is limited only by the availability of programs that will fit into

its internal memory. The user may personally program the computer to perform a particular task or may purchase a program from a software house.

All programs are known as **software** and the computer itself and the associated peripherals are known as **hardware**. Programs permanently stored in ROM are sometimes known as **firmware**.

The digital computer

The basic block diagram of a digital computer is shown in Fig. 2.1. The input and output devices are often known as **peripherals**.

Central processing unit

The **central processing unit** (CPU) is used to carry out all the processing as well as controlling the operation of the system. The CPU has two main parts: the **control unit** plus a number of registers, and the **arithmetic and logic unit** (ALU).

Control unit

The control unit controls the operation of the computer under the direction of the operating system (p. 65). It makes decisions regarding the handling of information: it controls the sequence of steps that are followed when a program is run; it controls the movement of data into and out of the ALU and tells the ALU which arithmetic, comparison, or logic operation it is next to carry out; it controls communication with the peripherals via the input/output circuitry. The program to be executed is stored in the internal main memory and the control unit **fetches** and **executes** the instructions in sequence. The **fetch—execute cycle** is the sequence of events that a computer follows to run a program. Each instruction in the program is in turn

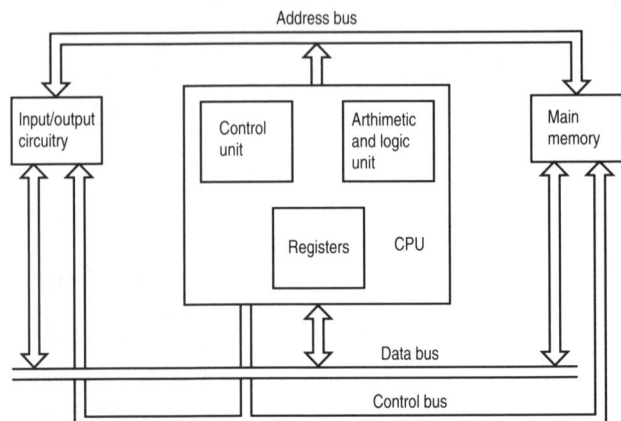

Fig. 2.1 Basic block diagram of a digital computer.

Start

Fetch on
instruction
from memory

Decode
instruction

Execute
instruction

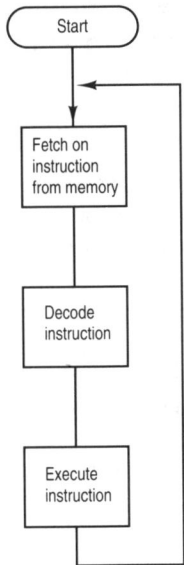

Fig. 2.2 Fetch-execute cycle.

individually fetched from memory and then executed. The FETCH part of the cycle refers to the operation by which the computer obtains the next instruction from memory. The computer first places the address of the memory location that holds the next instruction onto the address bus. It then sends a control signal to cause the READ action to take place and the data held in that memory location is then placed onto the data bus. The CPU reads and decodes the instruction, and then executes it. The EXECUTE part of the cycle refers to the computer carrying out the instruction. Execution may require no further data from memory, or another memory READ may be required in order to obtain more data upon which to operate. Once this data has been read into the CPU the instruction can be executed. The results of an instruction, and any data used, may be stored in the internal memory or, via an input/output circuit, placed in external memory. The fetch−execute cycle is shown by the flow chart given in Fig. 2.2.

Registers are required for the short-term storage of temporary variables. A register is a digital electronic circuit which is able to store a binary number. These registers temporarily store data as it is being moved from the main memory to the ALU or vice versa and also hold the intermediate results of calculations (see Chapter 7). To ensure that all the steps in a computer program are executed in the correct order all the operations within the computer are synchronized by an accurate **clock**. The control unit receives sequential instructions from the program being run, decodes each instruction in turn, and executes it. When a program has been run control is then transferred back to the operating system and this program (a) flashes the cursor on the screen, and (b) checks the keyboard to determine whether another key has been pressed.

Arithmetic and logic unit

All arithmetic and logic operations must take place using the **arithmetic and logic unit** (ALU). The ALU is also used as a temporary storage location that is used when moving data from one memory location to another. If an instruction involves the carrying out of an arithmetic, comparison, or logic operation on data then (a) the ALU must be instructed which operation to carry out and (b) the data must be transferred from memory to the ALU; the ALU will then carry out the instruction. The ALU is able to carry out arithmetic, comparison, and logic operations.

The arithmetic operations that an ALU is able to perform are addition, division, multiplication, and subtraction. These are all well known.

Comparison operations

There are several different kinds of comparison operation used in the execution of a computer program. These are as follows: greater than,

A > B; less than, A < B; equal to, A = B; less than or equal to, A ≤ B; greater than or equal to, A ≥ B; not equal to, A ≠ B.

LOGIC OPERATIONS
The logic operations are NOT, AND, OR, and exclusive OR.

- *The NOT function.* The NOT function merely inverses a binary number so that 0 becomes 1 and 1 becomes 0. This action is represented in binary algebra by placing a bar over a symbol; thus $\overline{A}$ means NOT A.
- *The AND function.* The AND function gives an output at binary 1 only when **all** the inputs are also equal to 1. If there are two inputs A and B the output will only be equal to 1 when both A and B are equal to 1. If either A or B is at 0 the output will be at 0. The Boolean equation that represents this action is A.B.
- *The OR function.* The OR function gives an output at binary 0 only when **all** the inputs are at 0. If there are two inputs A and B the output will only be at 0 when both A and B are at 0; if either or both inputs are at 1 the output will be at 1. The Boolean equation representing this operation is A + B.

Registers

The CPU contains several registers. An important part of the ALU is a register known as the **accumulator**. Data transferred from the main memory is first placed in the accumulator and is then processed according to the program instruction. The result of the operation is stored in the accumulator before being moved to a memory location. Some other registers are also provided to give temporary storage — e.g. to hold the operands.

The registers needed for program control are the **program counter** and the **instruction register**. The first instruction in a program is always held in a particular memory location and so the control unit always knows where to find it. The program counter initially holds the address of the first instruction to be executed. It is normally incremented by 1 during the execution of an instruction so that it always holds the address of the next instruction, but its contents might be modified by a **jump** instruction. The program counter is used to fetch the next instruction from the main memory. This instruction is loaded into the **instruction register** where it is first decoded and then carried out. Should an instruction involve any of the peripheral devices then that peripheral is sent a command by the control unit to operate it in the desired manner.

To move data between the CPU and the main memory two more registers are usually provided; these are the **memory address register** and the **memory buffer register**. The memory address register contains the address of the word in memory that is the subject of the next read, or write, operation. The memory buffer register is used

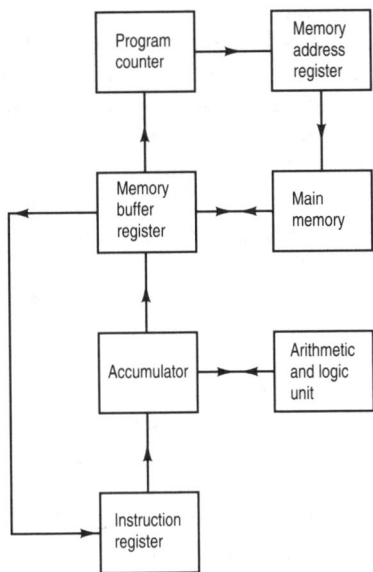

Fig. 2.3 Registers in a computer.

to store temporarily the data read from, or written to, the main memory. A read operation fetches the word into the memory buffer register where it is held until required by the ALU. In a write operation the CPU places the data to be stored into the memory buffer register. This data is then written into the memory independently of the CPU operations. The basic arrangement of the registers in a computer is shown in Fig. 2.3.

The sequence of operations for most instructions consists of a period of time known as the instruction cycle followed by another time period known as the execution cycle. At the beginning of the instruction cycle the contents of the program counter are transferred into the memory address register and that memory location is then accessed and its contents placed into the memory buffer register. This instruction is in two parts, one part, the **op-code**, contains the code for the operation that is to be carried out and the other part, the **operand**, contains the address of the data on which that operation is to be conducted. In the instruction 3E 79, for example, 3E is the op-code and 79 is the operand. The op-code is read out of the memory buffer register and is placed into the **instruction register**. This part of the instruction is decoded and then sent to the ALU to tell it which operation is to be performed. The operand is sent to the memory address register. The arithmetic/comparison/ logic operation specified by the instruction is then carried out and the result of the computation is transferred to the memory location whose address is now held in the address register. The CPU will then fetch the next program instruction from the main memory address held by the program counter and execute it. The operation is considered in some detail in Chapter 7.

Main memory

The internal main memory is provided by a mixture of semiconductor devices known as **random access memory** (RAM) and **read only memory** (ROM) with RAM being very much more common. Both types of semiconductor memory consist essentially of a matrix of memory cells, each of which is at a location that has a unique address (see Fig. 2.4). The location marked as A, for example, has the address

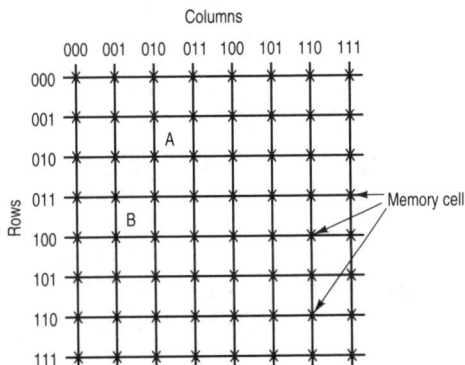

Fig. 2.4 Random access memory.

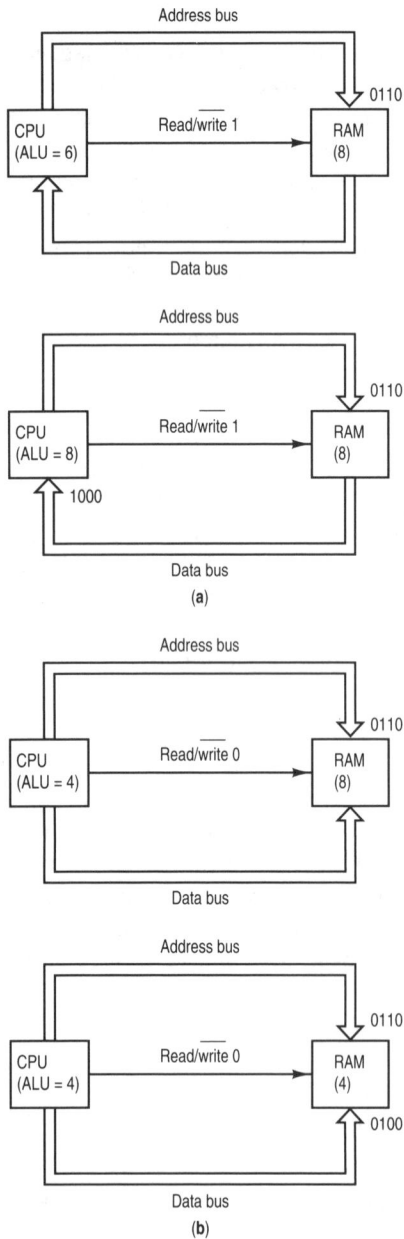

Address bus

Read/write 1

CPU
(ALU = 6)

RAM
(8)

0110

Data bus

Address bus

Read/write 1

CPU
(ALU = 8)

RAM
(8)

0110

1000

Data bus

(a)

Address bus

Read/write 0

CPU
(ALU = 4)

RAM
(8)

0110

Data bus

Address bus

Read/write 0

CPU
(ALU = 4)

RAM
(4)

0110

0100

Data bus

(b)

Fig. 2.5 (a) Reading data out of a memory location; (b) writing data into a memory location.

(column first) 010010, and location B has the address 001100. Any memory cell, or group of cells, can be addressed by the CPU and data either read out of, or written into, the memory. Reading data means that the data is transferred from the addressed memory location to the CPU. Writing data into memory means that the data is transferred from the CPU and stored at the addressed memory location. Depending upon the type of memory the read-out process may, or may not, destroy the data stored in an addressed location. If the read-out is destructive then it will be necessary for the data to be re-written into the memory immediately after read-out has taken place.

The **access time** of a memory is the time that elapses between the start of a read/write request and the completion of that request. The access time is the same for all locations in the internal memory and it is very short; this is why semiconductor memory is used to store the program(s) on which the computer is actually working at a given time. RAM is **volatile** and ROM is **non-volatile**; the term 'volatile' means that the stored data is lost when the power supply is removed. The main memory is expensive to provide and so its capacity is limited. Backing storage is therefore necessary to hold data and programs when they are not in current use.

The main memory is used to store input data as keyboard, or other, entry of data continues, and by the CPU to hold programs and data while a program is being executed. RAM is also employed to store the intermediate results of computations, the final results before they are outputted, details of memory locations used and what data is stored in which location, and so on.

Figure 2.5 shows the actions of reading data out of, and writing data into, a memory location. In figure (a) the read/write line is taken high and the address of the location placed on the address bus; for simplicity only a 4-bit address bus is shown. The addressed location is 0110. This memory location initially holds number 8 and the ALU holds number 6; after the contents of the location have been read the data bus has the binary number 1000 on it and the ALU also holds decimal number 8. The write action is illustrated in Fig. 2.5(b); initially the location holds number 8 and the ALU number 4 but the write action alters the contents of the location from 8 to 4.

The main (internal) memory of a computer is strictly organized and various parts of it are allocated to different tasks. Parts (a) and (b) in Fig. 2.6 show, respectively, the **memory map** for a small computer that is (a) running a word-processing program, and (b) running a BASIC program. The operating system of a computer is discussed in Chapter 4 but, briefly, it enables the computer to carry out all its functions. The stack is an area of short-term memory, often known as the **scratchpad memory**, that is particularly used in conjunction with subroutines. The bootstrap program is held in ROM and automatically loads the operating system when the computer is first switched on.

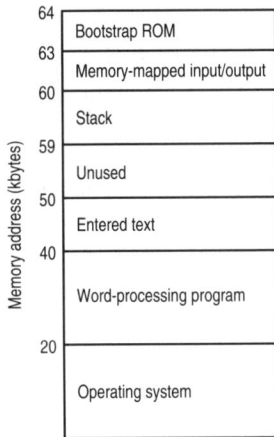

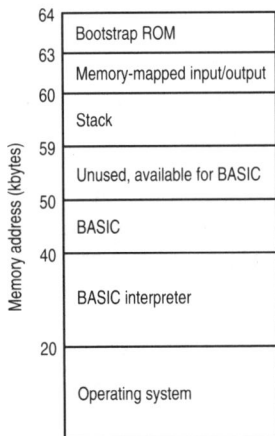

Fig. 2.6 Memory map of a computer; (a) when running a word processing program; (b) when running a BASIC program.

Data that must be retained at all times is stored in the long-term memory. Some of this may be provided by internally fitted ROM but most of it is provided by external memory. The **backing store** of a computer is used to hold data that is not required to be accessed immediately by the CPU. ROM may be used to hold the operating system, or a high-level programming language, in a small computer, but in the larger computers these programs are normally held on disk and loaded when required. All computers, however, must be fitted with a ROM that holds a **basic input/output system** (BIOS) program. The BIOS program performs test routines when the computer is first switched on, then loads the operating system from disk. The BIOS program also contains routines that access the system hardware. The external long-term memory is mainly provided by hard and floppy disk, but magnetic systems such as tape steamers are also employed.

Cache memory

The speed with which a CPU can process data is faster than the rate at which data can be supplied to it from the internal memory. To overcome the delay thus caused a high-speed **cache memory** can be employed. Data held in the main memory is pre-loaded into a much faster cache memory from which the processor is supplied with data. A cache controller predicts the instructions and data that will be needed next by the processor and copies them into the cache memory. If the controller guesses correctly, data transfer is considerably speeded up and this situation is known as a **cache hit**; if the controller guesses wrongly, the wanted information must be obtained from main memory and this is known as a **cache miss**. Typically, the cache hit rate is in the region of 80%. Cache memory is used in both mainframe and minicomputers as well as in the less expensive microcomputers.

Input/output circuitry

The function of the input/output circuitry is to transfer the signals appearing at the input terminals of the computer into the form required by the computer's internal circuitry, and vice versa. A keyboard is connected directly to the computer and it is not connected to the monitor. The computer may be programmed to 'echo' any key that is pressed so that each character is displayed on the screen to assist the operator.

Address, control, and data buses

A bus is a collection of conductors that are functionally related. There are three buses in a CPU: the **address bus**, the **control bus**, and the **data bus**.

Address bus

The address bus is a group of conductors over which a memory location (including memory-mapped input/output circuitry) is accessed. The address bus usually contains at least 16 conductors and it can then access up to $2^{16} = 65536$ different memory locations. The address bus is unidirectional from CPU to memory. The **directly addressable memory range** is the amount of memory locations that the CPU is able to access directly. It is determined by the width of the address bus.

Control bus

The control bus is a group of conductors that carries signals that synchronize the operation of the computer. Signals on the control bus may originate from either the CPU or from a peripheral.

Data bus

The data bus also consists of a number of conductors that are used to carry both instructions and data between the various parts of the computer. The data bus is bidirectional. The data width of the data bus is the number of bits that can be manipulated at the same time, which, for microcomputers, is generally either 16 or 32 but may be more in the larger computers. The data to be processed by a computer often has more bits than the data bus can handle. In such cases the data must be split up and stored in more than one location and then manipulated with more than one instruction. A large floating point number, for example, may require 32 bits and would need two locations in a 16-bit machine.

Types of computer

Computers are commonly classified into a number of groups depending on their size and processing speed, although the boundaries between these groups are becoming increasingly blurred. The groups are: mainframe computers, minicomputers, microcomputers or personal desktop computers, and portable computers. Any of these computers may be operated independently or may be linked to one, or more, other computers via a **data network**. A computer may be referred to as being either an 8-bit, or a 16-bit, or a 32-bit machine. This is a reference to the number of bits of data that can be moved over the data bus and processed at the same time. The larger the number of bits that can be handled at once the faster will be the processing speed of the computer. The speed is often expressed in mips (millions of instructions per second).

All computers must have a program, known as the **operating system**, that is either permanently installed or loaded automatically from a disk when the computer is first switched on or is reset. The functions of an operating system are:

- control and management of the input/output devices
- control and management of all processes within the computer
- management of the main memory
- management of all files and data
- organization of communications between computers in a network.

Mainframe computers

A **mainframe** computer is a physically large and expensive machine that has a very large working memory with a large backing storage and is able to support a large number of terminals and peripherals. The backing memory consists of both high-capacity hard disks and magnetic tape reels. A mainframe computer consists of several free-standing cabinets that contain the **central processing unit**, main memory, various items of backing storage, line printers, and a control console. A typical mainframe computer installation is shown in Fig. 2.7.

Each user gains access to the computer via a **terminal** that consists essentially of a keyboard and a monitor, collectively often called a **visual display unit** (VDU). Owing to its large internal memory and high-speed CPU, a mainframe computer is able to store and execute very large and complex programs rapidly. Its operation is so fast, typically 10 mips, that it is able to deal with a large number of terminals (apparently) simultaneously. The operating system controls the computer to go sequentially from one terminal to another to carry out whatever task each has for the computer; this process is known as **polling**. The main memory is provided by a RAM that is treated by the CPU as one continuous area of memory, any part of which

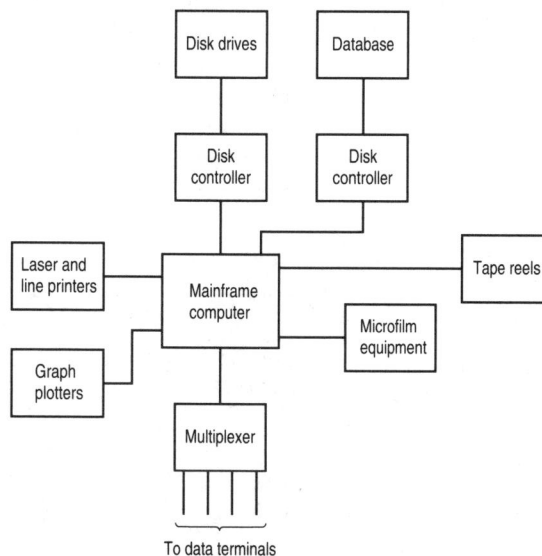

Fig. 2.7 Mainframe computer installation.

may be directly addressed. Because the CPU operates with such high speed, data cannot be read out, or written into, the RAM quickly enough to keep up with it and high-speed memory buffers — known as **cache memory** — are inserted between the CPU and the main memory. There may sometimes be several levels of caching between the CPU and the slowest elements in the computer system.

The CPU of a mainframe computer does not consist of a single integrated circuit (IC) but, instead, is built from several integrated circuit processors, each of which is allocated a dedicated task. This reduces the overall processing time because the processors are able to work concurrently in parallel rather than sequentially. The individual processors are often mounted on separate printed circuit boards. The ICs generally employ a technology known as **emitter coupled logic** (ECL) since this gives the fastest possible operation. The facilities provided by a mainframe computer may be needed at many different locations, some in the same building as the computer and others located some distance away. Distant terminals are linked to the computer by a connection in a data network. Data networks are considered in Chapter 11.

A mainframe computer is always provided with a very fast backing store. As this allows programs to be quickly loaded, there is not much need for internal permanent memory, which means that a mainframe computer does not have very much ROM other than a program to load the operating system when the computer is first switched on. The operating system and all other programs are loaded from a hard disk. This practice makes a mainframe computer very flexible with regard to the software employed.

Each addressable location in the main memory has a unique address and holds a number of bits of data known as a **word**. The number of bits is known as the **word length**, which may be fixed or variable. Many mainframe computers do not employ fixed word length memory locations since these can lead to some wastage of memory space. This point is illustrated in Table 2.1 in which it has been assumed that the characters CIRCLE are to be stored. In an 8-bit computer (only some home computers) each memory location holds just one character and hence six locations are required. A 16-bit computer can store two characters at each location and hence three memory locations are

Table 2.1

8-bit computer		16-bit computer		32-bit computer	
Location address	Contents	Location address	Contents	Location address	Contents
100	C	100	I C	100	C R I C
101	I	101	C R	101	E L
102	R	102	E L		
103	C				
104	L				
105	E				

necessary. A 32-bit computer is able to store four characters at each location and so just one memory location is used with a second that is only half filled. The situation could well be worse: a 32-bit location might be used to store a single character. Word lengths vary: 16, 24, 32, 36, and 64 bits have been employed in different computers. To overcome this difficulty some mainframe computers use variable length memory locations. Here a word has a length equal to one, or more, location widths and it is set to the required length each time data is transferred to or from the memory. In most cases each memory location is 8 bits in length. Program instructions may be of variable lengths and can all be transferred in one operation.

A common operating system used by some mainframe computers is known as **uniplexed information and computing system** (UNICS or more popularly UNIX). IBM mainframe computers include the system/370 range with **enterprise system architecture** (ESA) and the system/390 range (also known as ES/9000). These machines employ two operating systems, one known as multiple virtual storage (MVS) and the other as virtual machine (VM). MVS is used with large installations that carry out batch processing and VM is able to run other operating systems — for example, the IBM version of UNIX which is known as AIX.

Mainframe computers are employed for large data-handling applications, such as payrolls, customer accounts, weather forecasting, and airline booking systems.

Minicomputers

A minicomputer is a cheaper and smaller version of a mainframe computer that is able to support a smaller number of terminals (up to 40 usually), and is slower to operate. A computer of this type will probably have more programs stored in ROM since a minicomputer will be provided with less, and slower, backing storage. The main memory, provided by RAM, usually has a word length of 32 bits and is treated by the operating system as one continuous area.

This kind of computer employs several devices mounted in a single unit and will have at least 1 Mbyte of working memory which can be addressed as one continuous area. A minicomputer is capable of performing most of the tasks undertaken by a mainframe. Cache memory is commonly employed to speed up the operation of the machine. UNIX is the most popular operating system. Minicomputers are employed for small company business applications and for factory control systems.

Microcomputers or personal desktop computers

A microcomputer is a smaller and considerably cheaper machine than a minicomputer and it is able to support a single user. The cost of desktop microcomputers has fallen to such an extent that it is economical for businesses to provide one for the personal use of many of their staff, hence the term 'personal computer'. A personal computer is physically small enough to be mounted on a desk. There are two main groups of personal computer: IBM and IBM compatible machines known as **personal computers** (PCs), and Apple Macintosh machines. Since the IBM PC was introduced onto the market in 1981 many other manufacturers have produced compatible machines which are capable of running the same programs. Other small computers include the Amstrad PCW, the Amiga, the Atari, and Acorn BBC computers (although the Archimedes versions of the latter can be fitted with a PC emulator that enables the machine to run IBM compatible programs). In the remainder of this book all microcomputers will be referred to as PCs.

A PC has its CPU provided by a single integrated circuit (IC) known as a **microprocessor** and will have more programs held internally on ROM. The majority of PCs are either IBM machines or are IBM compatible and these machines are mostly based upon Intel microprocessor chips. Early models used the 8086 and 8088 chips which are identical except that the 8088 has an 8-bit, not 16-bit, data bus and external interface. Although some such machines are still in use, these ICs have been replaced with, in succession, the 80286, the 80386, and the 80486 microprocessors. The of these three computers is a 16-bit machine, the last two are 32-bit machines. The use of a 32-bit machine means that it is able to handle much bigger numbers than a 16-bit machine and, as a result, can address a much greater memory area. A more advanced chip, known as the Pentium, has now been introduced by Intel.

The operating system, DOS, used with most IBM and IBM compatible personal computers, was designed to run with the 8086 processor which is only able to address up to 1 Mbyte of memory. Of this, 640 kbytes is available for use by user-programs and the remainder is reserved for use by the system for tasks such as video memory, and BIOS ROM. The 80286 can access up to 16 Mbytes of memory using a 24-line address bus, but DOS only allows access to 640 kbytes. However, several Mbytes of 'expanded' memory can be made available, this extra memory is divided up into 16-kbyte blocks called **pages**. The CPU is unable to access this extra memory directly but it is able to switch these blocks, up to four at a time, into an area located at the top of the 640 kbyte ordinary memory. By rapidly exchanging different blocks of memory in and out of this area an effective increase in memory is obtained. The principle of expanded memory is shown in Fig. 2.8. The 80286 provides 'extended memory', which means that it is able to address 16 Mbyte of RAM but only in a special mode of operation that the DOS operating system does not understand. With DOS the 8086 and 8088 computers are also restricted to a direct memory access of 640 kbytes but they are able to use extended memory if a **DOS extender** is fitted or a newer

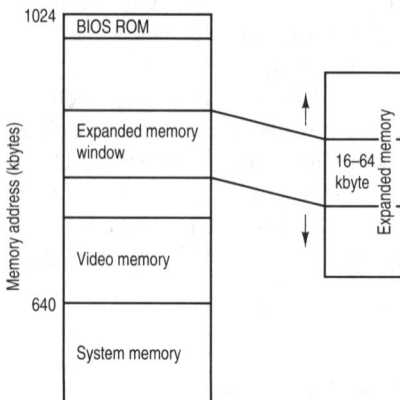

Fig. 2.8 Expanded memory.

Up to
16 Mbyte (max)
1024
BIOS ROM

Memory address (kbytes)

Video memory

640

System memory

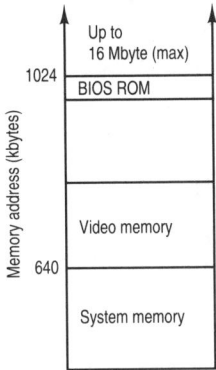

Fig. 2.9 Extended memory.

operating system such as Windows, OS/2, or UNIX is employed. The principle of extended memory is shown in Fig. 2.9.

The 8086/8 are the oldest microprocessors still used in PCs and they are mainly suited to such applications as word processors and simple databases. These devices were followed by the 80286 chip which is the minimum requirement if a commonly used graphical user interface known as Windows is to be employed. If faster operation is wanted and/or graphics are required, at least a 80386 PC is desirable. There are two versions of this machine, namely the 386SX and the 386DX. The DX employs 32-bit addressing and data transfer and is able to address up to 4 Gbytes of memory. The SX is also a 32-bit device internally but it addresses its backing storage over a 24-bit address bus, just like a 286 device. Hence it is able to address only 16 Mbytes of external memory. Because of the smaller external address bus the SX chip is cheaper than the DX alternative with only a small performance penalty. The 386SX chip is suitable for applications such as word processing, desktop publishing and CAD but better graphics results can be obtained if a 386DX computer is used instead. The most common chip now being employed in new PCs is the 80486, which also comes in SX and DX versions, and this gives a PC an excellent performance. The Pentium, like the 80486, supports a 32-bit instruction set, but is much faster.

Apple Macintosh microcomputers do not have the same problems with memory access because the 68000 family of microprocessors address all the fitted RAM as a single large block of memory.

Portable computers

There is an ever-increasing demand for a microcomputer that is physically small enough to be carried around by the person using it. A portable computer replicates the facilities and features of a desktop personal computer. Although there are several Apple Macintosh versions on the market the vast majority of portable computers are IBM PC compatible. Most portable computers are described as being either **laptop** or **notebook**. A notebook computer is about the size of an A4 sheet of paper and has a weight in the region of 4 to 8 pounds. Somewhat larger in dimensions are the laptop computers, and these are generally more convenient to use since both their screen and their keyboard are larger. The screen of all laptops and notebooks is mounted in a hinged lid that is opened up in order to use the computer. Most portables have provision for the device to operate with an external keyboard and screen and are battery operated. The battery can usually be recharged by plugging the computer into the mains electricity supply.

Portable computers have the same processing performance as desktop computers and even include a hard disk drive; however, the maximum RAM available may be limited.

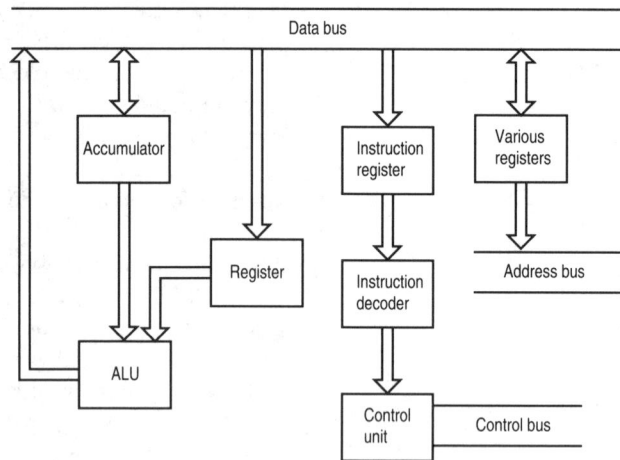

Fig. 2.10 Basic block diagram of a microprocessor.

Microprocessors will be considered in greater detail in Chapter 7 but the basic block diagram of one is shown in Fig. 2.10. The microprocessor can be seen to consist of control circuitry, an ALU, and some **registers** for short-term memory. Before a PC is able to run any software it must first be loaded with an operating system. The most commonly employed version operating system is the MicroSoft MS-DOS which, at the time of writing, is up to version 6.0. The working memory is provided by RAM and ranges from 1 Mbyte up to about 32 Mbytes. The larger the RAM the better since a PC is unable to run a program which requires more memory than the available RAM. PCs also contain a number of ROMs that may hold such programs as the operating system, a high-level language interpreter or compiler, and perhaps some other application programs.

A PC is provided with one (and sometimes two) floppy disk drive plus a hard disk drive (see p. 53).

Most of these PCs use the MS-DOS operating system but other operating systems that are available include: CDOS — concurrent disk operating system, a multi-tasking multi-user system; CP/M — control program for microcomputers, the original system for 8-bit computers and still used by Amstrad PCW machines; DR-DOS — Digital Research's version of DOS. UNIX is now being used in some of the more powerful PCs.

Other, non-IBM compatible PCs use other microprocessors; perhaps the main examples are the Apple Macintosh range that uses the Motorola chips M68000, M68030, and M698040, and the Amstrad PCW PC/word processors that use the Z80 chip.

Two PCs designed using the same microprocessor may have entirely different performances because of their different clock speeds. The faster the clock speed, measured in megahertz, the faster the PC will be able to perform a particular task. The processing speed depends also on the average number of clock cycles that occur per instruction.

Example 2.1

An 80246 PC has a clock frequency of 35 MHz and requires an average of 2.2 clock cycles to execute an instruction. Calculate the processing speed of the PC.

Solution

Processing speed $= (35 \times 10^6)/2.2 = 16$ mips. (*Ans.*)

Graphical user interface

A DOS provides a basic interface between a computer and its user. To simplify the operation of the computer from the user's point of view a **graphical user interface** (GUI) can be employed. A GUI uses icons to represent various programs and tasks that the computer has been set ready to perform; the user can move a pointer to point to the wanted icon, using mouse or keyboard, and then press a button or key to run the chosen item. One example of an icon is a picture of a filing cabinet. If this is selected the data is stored in a file. Another icon is a rubbish bin. If this is selected the information is discarded.

PCs have a wide range of commercial, domestic, and industrial applications, many of which were mentioned in Chapter 1. A PC can be used as a terminal for a mainframe computer if it is provided with a program called a **terminal emulator**. The PC will then act exactly like a VDU.

3 Input/output devices

Input devices

An input device is a means by which both data and programs can be supplied to a computer. The most commonly employed input device is a QWERTY keyboard which has the same keys and layout as a typewriter plus a number of special-purpose keys and function keys that are used to control and edit the display on the monitor. A numeric keypad may also be provided. Many other devices are also used to input data — for example, a mouse, a scanner, a document reader, a data link, and some kind of **transducer** that enables the computer to communicate with a controlled industrial equipment. As data is entered into a computer from a keyboard that data is visible on a monitor screen which allows the operator to view the data that has been entered. Most VDUs are not **intelligent**, that is they have no processing power of their own, nor do they include any storage facilities such as a disk drive. A skilled keyboard operator can only manage typing speeds of up to about 100 words per minute and this speed is much slower than the speed at which the computer is able to process data. In some cases the speed of inputting data can be increased by the use of **menus**. A menu gives, on-screen, a number of alternative actions and the operator is able to choose one of the actions by typing in a single letter, positioning the cursor over the wanted item and pressing the ENTER key, or the use of a mouse.

Mouse

A **mouse** is a hand-held device that is used in conjunction with on-screen menus and/or icons. As the mouse is moved around a pad or a desk, a pointer is simultaneously moved around the screen. The mouse has one or more buttons protruding from its top; these may be pressed to select the menu or icon to which the pointer is pointing. A mouse can therefore be used to select an icon, or to 'pull-down' a menu. An icon is a small picture which appears on-screen to indicate the facility it provides; for example, to file a document the user might point to the icon of a filing cabinet and when the button on the mouse is pressed the document will be filed; similarly, to discard an item the user might point to the icon of a rubbish bin.

Essentially a mouse is a mechanism that is able to detect movement in the backward/forward and left/right directions. There are three basic types of mouse: namely, mechanical, opto-mechanical, and optical.

1. **Mechanical**. A mechanical mouse has a small metal or rubber ball mounted so that it protrudes from the underside of the device and is in contact with the surface of the desk. As the mouse is moved over the desk surface the ball is rotated and its movements are converted by a transducer into electrical signals.
2. **Opto-mechanical**. This type of mouse is similar to the mechanical mouse but instead of using a mechanical transducer the movement of the ball turns some wheels that contain a number of blades. A light-emitting diode (LED) shines a light through each wheel and the emerging light is incident upon a light detector. As the wheels are turned they interrupt this light and the light detector converts the resulting pulses of light into electrical signals.
3. **Optical**. An optical mouse has no moving parts. It contains a pair of LEDs which direct light down to the desk top and some light detectors which measure the amount of light reflected from the desk. This type of mouse must be moved about the top of a special reflective mouse pad that has a number of coloured lines etched on its surface. As the mouse is moved around the reflected light is interrupted by the etched lines and the light detectors convert the interrupted light into electrical signals.

In all three cases movement of the mouse creates analogue electrical signals that indicate the direction in which the mouse is being moved. The analogue signals are converted into digital form and the digital signals are then processed by software to produce corresponding movements of the cursor on the screen.

Trackerball

A **trackerball** is a form of mouse that is used upside down so that the moving ball can be directed by hand. A trackerball is useful when there is little room on the desk top to allow a mouse to be moved around.

Scanners

A **scanner** is a device that is used to copy existing images into a computer. It is also used to read bar codes. An example of a scanner is commonly seen at the check-outs of many supermarkets. The operator passes the goods purchased, with the bar code downwards, over the scanner and a description of the goods is recorded. The **point-of-sale** (PoS) terminal also keeps a running total of the items bought and at the end of the transaction prints out a detailed list, with the cash offered and the change (if any). There are three basis types of scanner: the bar-code scanner, the **magnetic ink character reader** and the **optical character reader**.

Magnetic ink character reader

Special typefaces and magnetic ink are used to produce stylized characters on documents that can be read into a computer. Special character shapes are used to allow a magnetic detector to recognize each character without error. One of the main applications for **magnetic ink character recognition** (MICR) is in connection with the cheques that are issued by banks and building societies. Each cheque has numbers printed on it which indicate the cheque number, the customer's account number and the bank sorting code. Later, after the cheque has been presented for payment, the cash amount written on the cheque is typed alongside the account number using a magnetic ink machine. The MICR fonts are shown in Fig. 3.1. All the information needed to enable the bank's computer to adjust the customer's account can then be read from the cheque using a high-speed magnetic reader.

Fig. 3.1 MICR fonts.

Optical character reader

An optical character reader (OCR) optically scans text that has been printed on paper using a special typeface. The reader converts the scanned characters into ASCII signals that the computer is able to understand. The characters used, shown in Fig. 3.2, are fairly easy for a person to read. An OCR is relatively expensive and hence it is only used for large-scale data processing. OCR is more complicated and expensive than MICR and documents using it may be rejected by the optical reader if they have been folded or if there is any kind of unwanted mark on the paper. The system is particularly suited for use with large quantities of data recorded on pre-printed documents. The system is used, for example, by the electric and gas companies; bills printed out by the computer have a section that is returned by the customer along with the payment, and this section of the bill is read by an OCR to up-date the account.

Fig. 3.2 OCR characters.

Kimball tags

A Kimball tag is attached to each item of clothing in a shop. When the article is sold the tag is removed and is later scanned to produce data for input to a computer.

Touchscreens

A **touchscreen** is a method of inputting data to a computer by touching a part of the monitor's screen. Different touch-sensitive technologies are used but the basic principle of the capacitive method is as follows. The screen comprises a glass surface onto which a metallic oxide has been diffused. An electrode pattern is then formed in the oxide. The screen controller is able to measure the position of the capacitive coupling produced when the screen is touched by a finger or by a conductive pen. A user normally sees a menu of alternative actions and has merely to touch the one required. Software then converts the input signal that indicates the choice made into the form required by the computer. Touchscreens are often employed for applications where a computer is left unattended and is to be used by persons with no computer knowledge. An example of this is often to be seen in a shopping centre where customers are able to get information about goods and services by merely touching an on-screen icon with a finger.

Key-to-tape (or disk) systems

Whenever large quantities of data are to be keyed in by an operator the data can be recorded on either magnetic tape or a disk. The tape, or disk, can then be used at some later convenient time to input the data into the computer.

Point-of-sale terminal

Electronic point-of-sale (EPoS) terminals are commonly employed by department store 'Pay Here' counters and by supermarket checkouts. All the operator has to do is identify the item being purchased, enter its stock number and/or its price code (which may be done automatically using a scanner if the item is bar coded), and its description and price will be displayed on a screen. The terminal automatically totals the value of the goods bought and prints out a detailed receipt. The terminal is linked to a computer which uses the data to keep track of the items sold, of the remaining stock, and so on. The computer can use this data to automatically order new supplies for each shop as and when they are required. A large shop may have a computer on-site but smaller shops will not; their EPoS terminals will be connected by a data link to a computer that is sited elsewhere.

Telemetry

The collection of data from various points in a factory by remote means is common in industry. The data may be obtained from a keyboard having special keys, such as 'shut down', 'increase pressure', 'increase flow' and so on, or obtained from a transducer, such as a pressure gauge or a temperature sensor. Most transducers produce an analogue output signal which would need to be converted into digital form before being inputted to the computer. The data is transmitted over cable pairs to the computer where it is processed and, perhaps, used to control machinery.

Output devices

An output device is a means by which the results of a computer's computations can be made available to the end user — i.e. a human being or even another machine. The most common output devices are the monitor which displays the computer's output on the screen of a cathode ray tube, and the printer that gives a **hard copy** of the output. Other output devices may include a transducer of some kind that enables a computer to control some industrial process or machinery, a data link, or a graphical plotter.

Monitor

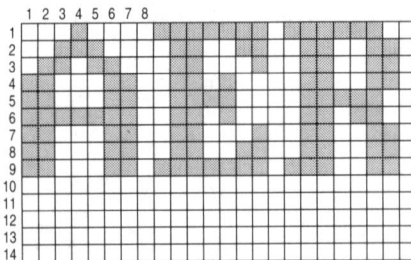

Fig. 3.3 Characters using a 8 × 14 pixel grid.

A monitor may give either a black-and-white (monochrome) or a coloured picture and it works in a similar way to a television receiver picture tube (p. 233). The screen is organized into a number of columns and rows of dot positions known as **pixels** and an image is built up using a combination of dots. Each character is generated in an 8 × 14 pixel grid (see Fig. 3.3). On a black-and-white screen each dot is either white or black (an absence of a dot), but for a coloured screen each dot is some combination of the three basic colours red, green, and blue (see p. 234). Screen resolution is quoted in terms of the number of dots measured first horizontally and then vertically. Thus, 640 × 480 denotes a screen that has 640 dots across the display and 480 dots from top to bottom. The product of the two gives the overall resolution, which, in this case, is 307 200 dots. If the total dots figure is divided by the area of the display, the dots per inch measurement is obtained. The higher the dots per inch figure the better will be the on-screen representation of text and graphics. Dot pitch is often quoted by manufacturers of monitors; this is the distance between adjacent dots, so the smaller the figure the better the resolution. Most SVGA monitors (see below) give a dot pitch of 0.28 mm.

The data to be displayed is held in the computer's video memory. The computer has to place each pixel on the screen and uses an IC known as a **video adaptor** to take the information from the video memory and convert it into the signals required to drive the monitor. The video adaptor scans the video memory taking the code for each character in turn and converts this code into a series of pulses that deflect the electron beam to 'draw' the characters. The **refresh rate** is the number of times per second that the monitor's display is refreshed by the video adaptor; the lower this rate the more likely it is that there will be noticeable flicker. Typically, the refresh rate is either 60 or 72 Hz.

Different quality displays are available; the basic display is the video graphics array (VGA). A VGA screen supports 640 × 480 pixels in any one of 16 colours. A super video graphics array (SVGA) gives various resolutions, with the most popular being either 800 × 600 or 1024 × 768 pixels in up to 16 different colours. An enhanced video graphics array (EVGA) gives 1024 × 768 pixels in up to 256 colours and can show up to 150% more information than a VGA display.

Very often the monitor and keyboard are combined, and are then collectively known as a **visual display unit** (VDU).

Printers

A printer is required to obtain a hard copy of the results of processing data. There are two main categories of printers: (a) impact types in which a character is transferred to paper by the impact of a type hammer pressing an inked ribbon against the paper, and (b) non-impact types. The principles of operation and the relative merits of the different kinds of printer used in offices are discussed in Chapter 5. Office printers are able to print one character at a time, whereas the printers used in computer centres can operate at very high speeds and (apparently) print one line at a time. They are known as **line printers**.

Plotters

A plotter is used to produce hard copy drawings and diagrams from a computer's output. Between 1 and 6 pens are moved onto the paper and are then controlled in their position by the computer to draw a required image. If required, each pen may use a different coloured ink.

When a peripheral is directly connected to a computer and is controlled by it, it is said to be **on-line.**

Memory or storage

There are five important requirements for a backing store:

1. It must have a large storage capacity.
2. Its cost per Mbyte stored must be less than for the main memory.
3. Its access time must not be too long, but it may be considerably longer than for the main memory.
4. Information need only be written in, or read out, as blocks of words, such as files.
5. It must be non-volatile, i.e. be able to retain its data when the power supplies are turned off.

The backing store is provided externally to the computer primarily by hard and/or floppy disks but also by magnetic tape and the read-out is non-destructive. Tape storage methods record the data serially. When required, existing data can be over-written by new data. The access time of backing stores is longer than for a semiconductor memory but they are non-volatile. The backing store is used for

(a) data and programs that need to be stored for long periods of time;
(b) programmes and data that are being processed by the computer but for which there is not sufficient room in the internal memory.

Information stored in a backing store can be stored in a place away from the computer until it is needed. It may also be taken away and used in another computer.

The capacity of a memory is measured in terms of the number of bits (or sometime bytes, where 1 byte = 8 bits) of data it is able to store. The capacity is always equal to some power of 2 and some typical figures are as follows (k = 2^{10} = 1024; M = 2^{20} = 1.0486 $\times 10^6$):

- RAM and ROM: 16 k, 64 k, 256 k, 1 M, and 4 M (in bits). In a PC the basic RAM provided can always be upgraded (expanded) by the fitting of additional chips and capacities of up to 32 Mbytes are common.
- Floppy disk: 720 k and 1.44 M (in bytes).
- Hard disk: 44 M, 106 M, 130 M, 210 M, and 245 M (in bytes).

The most important characteristics of a backing store are its latency time, its capacity, and its transfer rate.

1. **Latency time** is the time that elapses between the receipt of an access signal by the store and the start of a data transfer from the store to the computer. Closely associated with latency time are the **read time**, which is the time taken from start to completion for a data transfer, and the **propagation time** which is the time that elapses from the instant of executing an instruction to the instant the access signal arrives at the store. The **access time** is the time taken to retrieve data from the backing store; it is equal to the sum of the propagation time, the latency time, and the read time. It varies from about 12 ms to about 80 ms.
2. The **capacity** of a backing store is the amount of data it can hold. The capacity is usually quoted in **bytes**, where 1 byte is equal to 8 bits.
3. The **transfer rate** of a backing store is the rate at which it is capable of transferring data between itself and the computer. A typical transfer rate is 170 kbyte/s.

Data is stored, and accessed, on a disk in blocks. This method of storage is less flexible than the storage of, and access to, data stored in the internal memory which is random. If a program requires an alteration to just one bit of data that is stored on a disk then a whole block, perhaps as large as 512 bytes, must be read out, amended, and then written back to the disk.

Cache memory

Modern CPUs need data at a quicker rate than a disk controller can allow. Normally, when a file is opened the computer reads it into internal memory from a disk and then the data is taken from memory as required. If a **cache memory** is employed a file is read into the cache and the next time the computer wants to read data from the file it will take it from the cache instead of from the disk. This enables the CPU to access the data in a much shorter time, typically nanoseconds instead of milliseconds, and so gives an overall improvement in speed. Cache memory is provided with many disk interfaces and it may be provided by either hardware or software means; the former method has the advantage that it does not use up any of the internal memory locations. The basic block diagram of the arrangement of a cache memory is shown in Fig. 3.4.

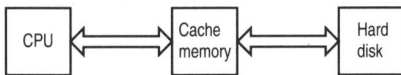

Fig. 3.4 Cache memory.

Floppy disks

A floppy disk drive is commonly employed with both personal and home computers. A floppy disk is a thin flexible plastic disk that has been coated on both sides by a magnetic material and is enclosed within a protective cover. The disks are usually of either 3.5 in. or 5.25 in. diameter although a few computers (the Amstrad PCW, for example) use 3 in. disks. Figures 3.5(a) and (b) show the appearance of the 5.25 in. and 3.5 in. floppy disks respectively. The disk cover contains a small slot through which the magnetic surface may be contacted by the disk drive's read/write head(s). Both 5.25 in. and 3.5 in. disks have a plastic cover and the 3.5 in. disk has a metal shutter that protects the magnetic surface whenever the disk is not in a disk drive. A read/write head consists of coils of wire mounted in a tiny case that is moved across the surface of the disk. The head reads information from the disk by sensing magnetic variations, or writes information into the disk by using electrical pulses to create magnetic areas on the disk's surface. There is only the one read/write head for each surface of the disk that is used, i.e. either one or two heads are provided. When a disk is inserted into a drive it is rotated at high speed and a read/write head is brought into contact with its magnetized surface. The head is moved across the disk to be positioned on the surface of the required part of the disk to either read the data stored at that point, or write new data. The disk drive mechanism is controlled by the computer, via a disk interface IC. When a floppy disk is inserted into a disk drive a hub engages the central hole in the disk and rotates the disk at a speed of about 360 revolutions per minute. The protective plastic cover has slots cut in it so that the disk drive's read/write heads are able to make contact with the magnetic surfaces on both sides of the disk. By moving the head in and out of the drive it is able to make contact with all parts of the disk's surface. When a head is

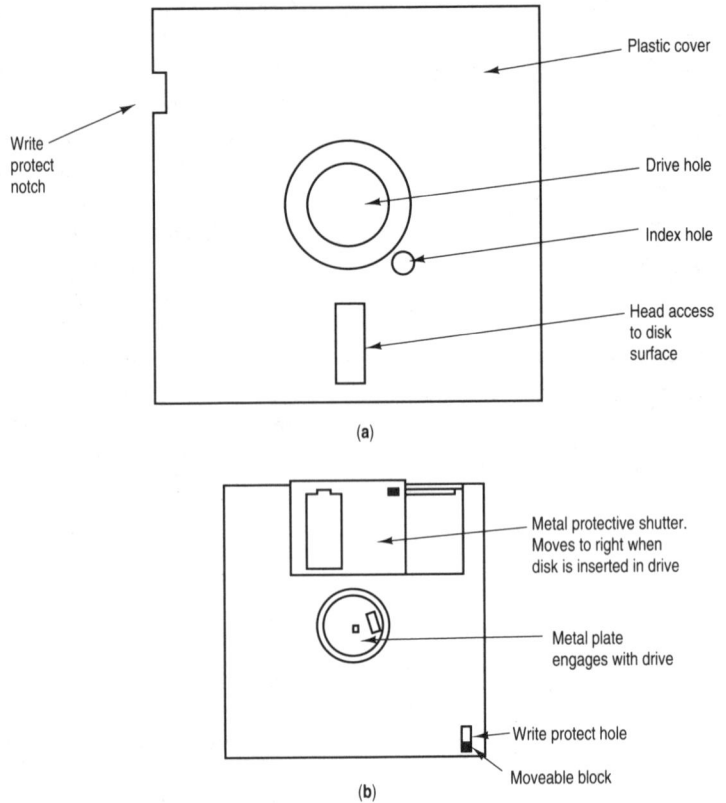

Fig. 3.5 Floppy disks: (a) 5.25 in. and (b) 3.5 in.

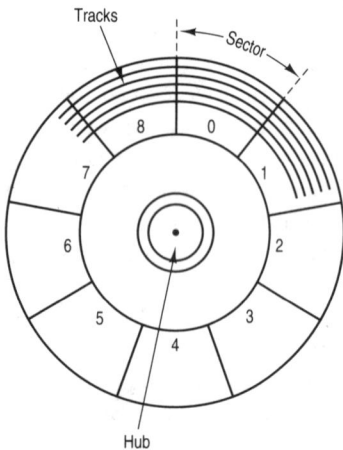

Fig. 3.6 Floppy disk; tracks and sectors.

writing data onto a disk pulses of electricity are passed through a coil of wire to set up a magnetic field that magnetizes the part of the disk surface directly underneath the head. When data is read from the disk the changing magnetism of the disk as it rotates induces alternating voltages into the read coil.

A floppy disk has a number, either 40 or 80, of concentric recording tracks on which information is stored; and most systems use the 80 track format since it offers twice the capacity. Each of these tracks is divided into a number (usually 9) of sectors or **blocks**, each of which is able to store up to 512 bytes of data. This is shown in Fig. 3.6. A sector is the smallest area of disk storage space that can be read from, or written to, in one operation. Each sector is identified by an address label located at its start. Sectors are used for the addressing of areas of the disk for read/write purposes and are all of the same size on a particular disk. The same amount of data can be stored on each track within a block. Each sector is given a unique address that enables the disk drive to locate any stored data. Data is written into, or read from, one block at a time. The **packing density**, or **recording density**, in bits/cm, is the amount of data that can be stored per centimetre of track. To ensure that the amount of data stored in each track within a sector is the same, a higher packing density is necessary for tracks near the centre of the disk than for tracks near the outer edge.

Many floppy disks are double sided and others are also double density. The term 'density' refers to how closely data can be stored on the disk. A high-density floppy disk can store 1.44 Mbytes of data. Floppy disks are relatively cheap and provide direct access storage of data; this means that the disk drive is able to gain immediate access to any file on the disk. To write new data onto a floppy disk, or to read data from a disk, the disk must be placed into a disk drive. In most cases this drive is an integral part of the computer but may also be a separate piece of equipment. Many computers have two floppy disk drives and these are usually labelled A and B.

Example 3.1

Calculate the storage capacity of a 3.5 in. floppy disk.

Solution
If there are 9 sectors and 80 tracks on each surface then

$$\text{Capacity} = 2 \times 80 \times 9 \times 512 = 737.3 \text{ kbytes.}$$

In practice, this figure is reduced to 720 kbytes. (*Ans.*)

Note that high-density types have double this capacity, i.e. 1.44 Mbytes.

Formatting

A new floppy disk is completely blank and before it can be used it must first be **formatted**. This means that the disk must be magnetically configured to work with a particular computer. After formatting a disk has certain data written onto it that specifies the sectors, produces a file directory, a file allocation table, and a **boot sector**. The file directory keeps a record of each file written onto the disk, and the file allocation table keeps a record of all the sectors used and to which files they belong. The boot sector is read when the computer is first switched on, or is reset, and causes the computer to automatically run a specified program from the disk, e.g. the operating system.

The capacity of an unformatted 80 track 5.25 in. floppy disk is: single-sided single density, 250 kbytes; single-sided double density, 500 kbytes; double-sided single density, 500 kbytes; double-sided double density, 1 Mbyte. The high-density version gives 1.2 Mbytes. For 40 track disks the capacity is 50% less. A 3.5 in. disk has an unformatted capacity of 500 kbytes for a single-sided disk and 1 Mbyte for a double-sided disk. High-density types have a capacity of 1.44 Mbytes. Lastly, a 3 in. disk has an unformatted density of 720 kbytes.

Floppy disks can be used:

(a) as backing storage for PCs
(b) to supply software for PCs and home computers
(c) to collect and store data for transfer to another system

Hard disks

A hard disk is a circular aluminium, or glass, plate whose two surfaces are coated with a magnetic material. A hard disk has a much higher capacity than a floppy disk and a very small access time but, on the other hand, it is much more expensive. The higher the capacity of a hard disk the more expensive it is. There are three types of hard disk drives in use: drives with moving read/write heads, drives with fixed heads, and Winchester drives. The movable head type of disk drive is often used to provide backing storage for mainframe computers. These hard disks may be either single exchangeable or provided in a disk pack. The exchangeable type are of varied diameters ranging from 3 in. up to 18 in., and they are inserted into the disk drive as and when required. A single exchangeable disk is sometimes called a **cartridge disk**. The hard disk pack is a set of hard discs that are held parallel to one another by a spindle (see Fig. 3.7). The disk drive mechanism consists of a number of arms at the end of which a read/write head is mounted. The read/write heads pass between the individual disks, one per surface, and cannot move independently of each other. All the disk surfaces except the outer two are coated with a magnetic material and are able to store data. As with floppy disks, each surface is formatted to have a set of concentric tracks, typically between 200 and 800, and each track is divided into a number (typically 8) of blocks or sectors. When a disk pack is inserted into a disk drive the read/write heads are very closely positioned over the first track but do not touch it. The disks are rotated at a very high speed of several thousand revolutions per minute (typically 2400−3600). The arms move in and out across the disk to access the wanted track and access time is typically 25−100 ms. A disk pack can be removed from the drive and replaced with another pack.

The fixed head type of hard disk drive has a read/write head for every track of every surface on each disk (see Fig. 3.8). The disk pack is often not visible and since the heads do not have to move to access data the access time is very small, typically 10−40 ms. This type of disk drive is more expensive than the movable head type.

Personal computers employ a Winchester hard disk drive that uses a single moving head and a single 2.5, 3.5, or 5.25 in. hard disk inside a hermetically sealed unit (see Fig. 3.9). 5.25 in. drives are now only used for very high capacity units and 2.5 in. drives are used for laptop computers. 1.8 in. drives are starting to be employed in both laptop and notebook computers. The read/write heads rest on the disk when

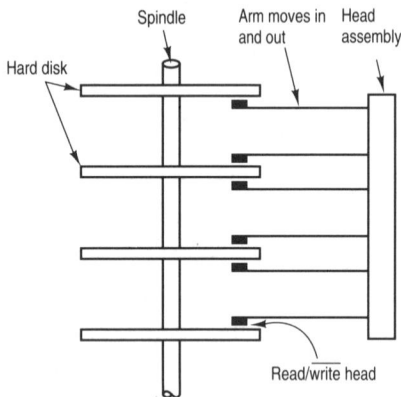

Fig. 3.7 Movable head hard disk drive.

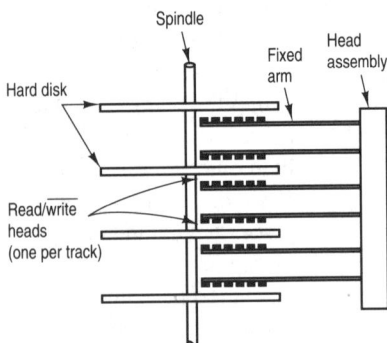

Fig. 3.8 Fixed head hard disk drive.

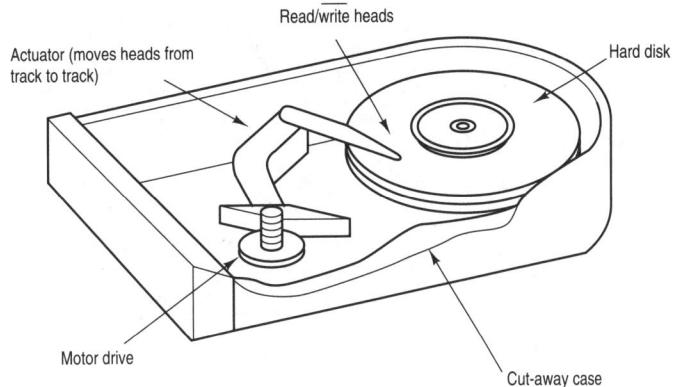

Fig. 3.9 Winchester disk drive.

it is stationary but are held just above it when the disk is rotating. This practice allows the disk to be rotated at a much faster rate than can be achieved with a floppy disk. Consequently, both the number of tracks and the recording density per track can be considerably increased. Typically there may be as many as 500 tracks with each sector holding 512 kbytes of data. Since the disk is enclosed within a sealed disk drive unit it cannot be exchanged like a floppy disk and it is good practice to regularly back-up its contents onto either a number of floppy disks or onto a magnetic tape system. Hard disk drives for PCs are supplied in a number of different capacities, such as 40/42/44, 85, 99, 105/6, 120, 130, 205, 210 and 330 Mbytes. Winchester hard disk drives and disks are cheaper than the types used in conjunction with mainframe computers and minicomputers, and they are also more reliable.

There are two main types of Winchester disk drive controller in use: these are known as **integrated drive electronics** (IDE), and **small computer system interface** (SCSI). IDE drives are both cheaper and simpler to provide since the disk controller is an integral part of the drive. SCSI drives use an interface that may also be employed to connect other peripherals, such as printers and tape streamers, to a PC. Increasingly, PCs use the IDE controller since it is both cheaper and more reliable unless a very large storage capacity is required, in which case SCSI is employed.

Magnetic tape storage

Magnetic tape is often used for long-term serial access backing storage for a mainframe computer or a minicomputer. The plastic tape employed is coated with a magnetic material and is supplied in reels that are 1.25 cm wide and some hundreds of metres in length. A tape is generally kept within a cartridge. The width of the tape is divided into a number (usually 9) of tracks. A reel-to-reel tape deck system,

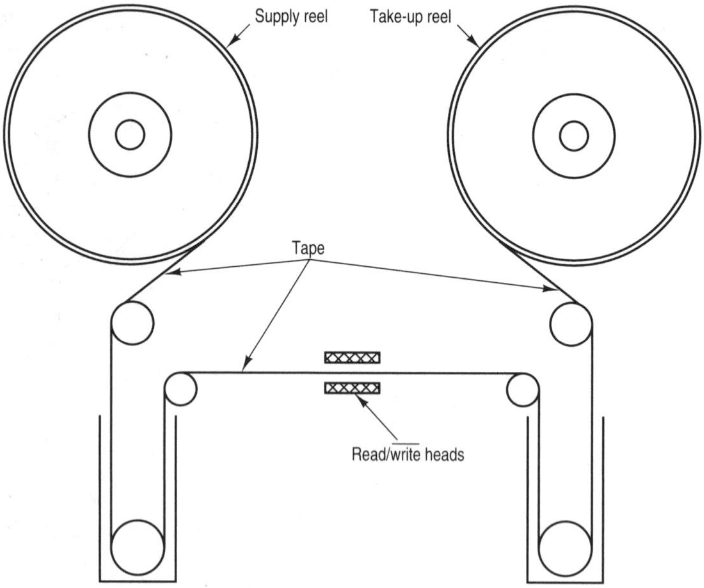

Fig. 3.10 Reel-to-reel tape storage.

like that shown in Fig. 3.10, may be employed. The tape is fed from the supply reel, past the read/write heads, onto the take-up reel. There is a read/write head for every track on the tape. Because the data is stored sequentially it can take some time for particular data to be accessed and so the system is not well suited for fast loading. Typically, a reel of magnetic tape is able to store 10 Mbytes of data.

A **tape streamer** is a small tape drive that is used for back-up, for archival storage and for data interchange. When files are up-dated on a hard disk they can be copied on to a tape streamer at very high speed. This practice avoids the cost of using expensive hard disks for back-up, or the need to deal with a large number of floppy disks for this purpose, which would otherwise be necessary. A tape streamer is a continuous loop of tape that is held within a cassette and is able to accept a continuous stream of data at a very high transfer rate. One such tape will have sufficient capacity to store all the data held by most hard disks and the data can be loaded back to the disk much more quickly than from several floppy disks. Tape streamer provides a relatively cheap method of long-term storage and is often used in conjunction with PCs. For *very* long-term storage of data a microfilm or microfiche system may be used (p. 13).

CD-ROM disks

A CD-ROM disk is an aluminium-coated plastic disk of about 5 in. diameter. During manufacture of the disk a laser beam is used to place a digital dot pattern, consisting of very small pits in the metal surface, in a single spiral track. Data is stored by means of these pits; the presence of a pit represents binary 1 and the absence of a pit represents binary 0. A CD-ROM has a very high storage capacity, typically 600 Mbytes, but it is a read-only device. Data is read from the disk by moving a laser beam radially across it as it rotates. This laser beam is of lower power than the one used in manufacture to ensure that no other pits are created. The laser beam is reflected back from the disk surface at each point where it is incident on a pit-free area, but each time a pit is encountered the light is scattered and so much less light is reflected back. The reflected light beam is focused onto a light detector which converts the varying reflected light intensities into an analogue electrical signal. In turn, this analogue signal is converted into the digital information required by the computer.

4 Software

Before a computer is able to perform a given task it must first be loaded with a **program**. A program is a sequence of instructions that specifies the various operations that the computer must carry out on data to achieve a required result. The computer will execute each instruction in the program in the order in which it is given unless some form of **jump** instruction is reached. An **algorithm** is the name given to the steps that must be followed by the program for a particular task to be carried out, or for a problem to be solved. If the algorithm is written in a language that the computer is able to understand then that algorithm is represented by a program. Before a program can be written the task to be performed must be broken down into a number of small steps aided by the use of either a flow chart or a pseudocode.

Computer programs are known as **software** unless they are permanently held in a semiconductor memory device known as ROM when they are called **firmware**. There are three main classifications of software: (i) systems software, (ii) support or utility software, and (iii) application software. System software enables the computer to work and to interface with its peripheral devices. Since they are constantly needed systems programs are held in the internal memory while the computer is working. Support (utility) programs perform functions that are only occasionally required and these programs are held on disk and are loaded when needed. Application programs are the software that makes the computer perform a particular task such as acting as a word processor.

Each operation performed by a computer is in response to instructions that are in the form of binary signals, i.e. combinations of logic 1 and logic 0. Each instruction tells the computer to do something specific; for example, to add two numbers together or to store another number. It is very difficult to write a program to instruct a computer what to do using binary notation since the process is very error prone, the program will be very difficult to understand (since it consists of long streams of 1s and 0s), and it is difficult to keep track of the addresses of the memory locations that have been used. The binary instructions are said to be in **machine code language** because that is all the computer can understand. If the instructions are written in any other language the program will have to be translated into machine code before the computer is able to execute the instructions. Once a program has been executed, control of the computer is taken over by a systems program that flashes the cursor onto the screen and scans the keyboard waiting for a key to be pressed.

Each item of data that is handled by a computer may be either a constant or a variable. Constants, which may be either numeric or a **string** of letters, do not change in value while the program is running; examples are many and may include the number of litres per gallon, a person's National Insurance number, the name of a business, the address of a tax office, and so on. A variable is a piece of data whose value may vary at any time during the running of the program, such as the price of petrol per gallon.

Many programs are **menu-driven**. This means that the user is required to select one of a number of possible choices that are displayed in an on-screen menu. Selection of a particular choice is made by entering at the keyboard the number of that choice, or by using a mouse to move a pointer on-screen until it is over the required choice and then pressing a button on the mouse.

A **window** is a section of the screen which can be dealt with separately from the remainder of the screen. A window may display a menu of choices while the remainder of the screen shows the computing activity that is in progress. **Icons** are generally used instead of words; examples of icons include an office filing cabinet to indicate a file, a waste-paper basket to indicate deletion, or a printer to indicate print.

Low-level languages

A computer language may be either **low level** or it may be **high level**. A low-level language is written in either machine code or an **assembly language**. Writing a program in machine code is a very tedious and extremely error-prone process. The various numerical operation codes bear no obvious relationship to their functions and the address of each memory location used must be carefully noted and its numerical value used whenever that location is referred to. For example, the binary coded number 1000 1001 could represent the instruction ADD with carry, 1000 1000 could mean ADD, and 1000 0000 could mean subtract. Any changes to a machine code program which alter the numerical address of any memory location require careful checking and the alteration of any references to that location.

The writing of a low-level program can be simplified if more easily remembered groups of letters, known as **mnemonics**, are used to replace the groups of binary numbers that represent the instructions that are actually operated on by the CPU, and labels (also known as **symbolic addresses**) are used to indicate the address of each memory location used. The mnemonic code is generally a set of abbreviations designed to be fairly easily interpreted. For example, if, for a particular computer, some binary digits or **bits** instruct the computer to ADD two numbers together, the mnemonic used could (obviously) be ADD. Also the address in which the sum of the addition is to be stored can be given a name, e.g. SUM. The mnemonic often represents an instruction to act upon the data stored in the next memory location. A low-level instruction has two parts: the op-code and the operand. The **op-code** is the part that specifies the function of the instruction, i.e. it tells the CPU what it is to do with the following data; and the **operand** is the part that specifies either the value of the data or the address of the memory location in which the data is held. The list of mnemonics that a CPU is able to understand is given by the **instruction set** for that processor. An instruction set is designed for use with a particular CPU. Three typical examples are: ADC means add with carry, LDA A means load accumulator A, and STA A means store the contents of accumulator A. Other mnemonics are given later. Each mnemonic corresponds to just one machine code instruction.

The use of a low-level language instead of machine code introduces the problem that the computer is unable to understand the instructions. It is therefore necessary to employ an **assembler program** that is able to convert the low-level language program fed into the computer into the machine code that the computer is able to act upon. The assembler also decides where the source program and the data are to be stored, keeps a note of the symbolic addresses used, and detects any errors. The assembler program is either held in a ROM or on a disk; in the latter case it must be loaded into the computer immediately the machine is switched on. The basic concept of an assembler is shown in Fig. 4.1. The assembly-language program to be translated is known as the **source code** and the translated machine code is known as the **object code**.

A low-level language is not 'user-friendly' and it takes some time for a user to become familiar with it. Another disadvantage of a low-level language is that, because of its relatively simple instruction set, it may take several instructions to perform even a simple task. An assembly language is appropriate only for the CPU for which it was written and is hence said to be **machine orientated**.

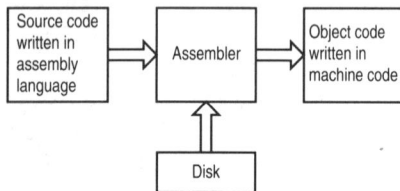

Fig. 4.1 Use of an assembler.

High-level languages

In an attempt to make the programming of a computer as easy as possible a number of **high-level languages** have been developed. A high-level language has an operation code set that contains words that are often close to English. This makes the program easier to write, to understand, and to modify. Each statement (single instruction) is translated into several machine code instructions by a translation program. This means that the language is less efficient, i.e. slower and uses more memory, than a low-level language. A high-level language has the advantage that it has been designed to be usable on a variety of different machines, i.e. it is **portable**. Before the computer can execute a program the high-level language must be translated into machine code by a translation program that is known as either an **interpreter** or a **compiler**. A high-level language is problem orientated rather than machine orientated, this means that the programmer does not need any knowledge of the operation of the computer. He must just follow the rules of the language that is employed.

When a program is to be written the first step should be to analyse the problem. Next, the types of inputs, which data is to be stored on disk (or tape), which calculations are to be carried out, and how results are to be outputted, must all be specified. An algorithm of the problem is then generally produced, using either a flow chart or pseudocode, and this algorithm is used to assist in the writing of the program. The compiler (or interpreter) is then used to detect any errors such as spelling or omissions before the program is tested by running it using test data.

Relative merits of high-level and low-level languages

1. High-level languages are easier to write, to modify and to understand.
2. High-level languages are portable.
3. Low-level languages are more efficient, i.e. they run faster and use less memory space. Obviously, the more memory that is occupied by a program the smaller is the amount of memory available for data.
4. Low-level languages allow programs to be written that make full use of the capabilities of the computer.

Most programs are written in a high-level language unless: the utmost efficiency is required, e.g. systems programs; the program is to be stored in a ROM; an interactive graphics program is written, like a computer action game; the program must be executed in the minimum amount of time, i.e. real-time operation, which is often the case for industrial control, military, and telecommunications applications.

Types of data

The data handled by a computer may be numeric, alpha-numeric or logical, and may be either constant or variable. Numeric data may be either integer or real. Integer numeric data consists of positive and negative whole numbers and zero, e.g. 0, ± 2, ± 7. Integer numbers are used wherever possible since they can be stored exactly and operated upon quickly. Real numeric data consists of all the numbers that the computer is able to handle; it includes the integer numbers plus all the fractional numbers in between, e.g. 0, 1.5 and 7.9. The usual way in which real numbers are handled in a computer is known as **floating point representation**. With this method numbers are written as a number in the range 0.1 to 1 times a power of 10. Thus, $5.5 = 0.55 \times 10^1$, $55 = 0.55 \times 10^2$, $555 = 0.55 \times 10^3$, $0.55 = 0.55 \times 10^0$, $0.055 = 0.55 \times 10^{-1}$, and $0.0055 = 0.55 \times 10^{-2}$. Here 0.55 is the **mantissa**, 10 is the **base** or **radix**, and the power of 10 is the **exponent**. The exponent is always an integer number.

Alphanumeric data, often known as **strings**, includes the letters of the alphabet, the digits 0 through to 9, and the various punctuation symbols such as :, ?, and !. Most computers use a character set specified by the **American Standard Code for Information Interchange** (ASCII). In the ASCII code each alphanumeric character is represented by a particular combination of binary 0s and 1s. Table 4.1 gives eight examples of ASCII character codes.

Table 4.1

A	01000001	T	01010100	a	01100001	t	01110100
5	00110101	%	00100101	(	00101000	?	0011111

Boolean data can have any one of two different values: a data value is either TRUE or FALSE.

Software

Software is broadly classified into one of three categories: namely, **systems** software, **support (utility)** software, and **application** software.

Systems software

Systems software is provided with a computer to control the performance of the system and to provide certain often-used facilities; it is usually supplied by the computer manufacturer on a disk. The most important systems programs are the **basic input/output system** and the **operating system**.

Basic input/output system

The **basic input/output system** (BIOS) is a collection of programs held in a ROM. When the computer is first switched on, a program in BIOS — known as **power-on self-test** (POST) — starts to run and checks that all parts of the computer are working correctly. If a fault is found an error message is given and (usually) the computer will not boot up. After POST, BIOS runs a program that loads the operating system from disk. Once the operating system has been loaded, BIOS provides several services which assist in keeping the computer running, including interfacing the computer to its various peripherals, such as the keyboard, the monitor, a printer, and the disk drive(s). A floppy disk managed by BIOS is always labelled A (and B if there are two) and the first hard disk is labelled C. When the operating system has been loaded a prompt appears on the screen and the user is then able to start running an application program.

Operating system

The functions of an operating system are many and include the following:

(a) it manages and controls the execution of application programs
(b) it manages the inputting and outputting of information
(c) it manages files stored in the backing store
(d) it manages peripherals
(e) it carries out the user's instructions
(f) it indicates any errors in the operation of the peripherals
(g) it activates support (utility) programs
(h) for large computers, it implements multi-tasking and multi-user operation.

An operating system contains a number of smaller programs each of which performs a particular function, such as reading the data entered at the keyboard, running programs, and controlling graphics. A part of the software known as the **disk operating system** (DOS) allows the user to perform such tasks as listing the contents of a disk, deleting or up-dating files, and interfacing the disk drive to the computer. When a program or data is saved to disk the operating system allocates the track and sector to be used (unless IDE is employed, see p. 57) and keeps a record of which sector has been used for which file. A file is found by its name but its contents are not legible, an application program must be used that can load, interpret and display a file. The operating system includes a number of commands which the user can use to instruct the computer to perform various actions; examples include: list all the files held on a floppy disk, select a high-level language compiler, access the editor program (p. 70) and activate a support program.

While an application program is running, the operating system manages and controls the execution of the program and it is ready to take over if the program is about to disorganize the system, perhaps by trying to use memory locations that have been reserved for another purpose. It also controls, along with BIOS, the input and output of information and manages all the peripheral devices. Any operation that is repeatedly needed by an application program should be a part of the operating system. Multi-tasking, which allows a computer to process more than one task at the same time, is provided by some operating systems.

In a large computer the operating system also controls communication with several terminals at the same time, ensures that individual programs run without mutual interference, keeps a log of the programs run, maintains security, i.e. ensures that all users log in and log out and give the correct identification and password before being given access to files, and organizes the use of memory, which has to be shared between different users at the same time. If the users pay for the computer time used, the system keeps a check on that time and generates accounts of the monies owed.

An operating system is usually stored on a disk and is automatically loaded into the computer immediately the machine is switched on. The system disk can then be removed from the disk drive. Together, the BIOS and an operating system program — IO.SYS in the PC MS-DOS operating system — load the remaining system programs into the internal memory. The operating system is a complex program and it is written by a specialist software firm. Very often the system disk also includes some application programs. If an operating system is portable the software can be used with different computers. This is advantageous since it means that programmers and users may use different machines without having to learn the intricacies of each different operating system. An operating system is made portable by either writing it in a high-level language, writing it in C, or by writing it in the assembly language of a popular microprocessor that is used by several different computers.

Single-user systems

An operating system may be either a single-user, a multi-user, or a distributed system. A single-user operating system organizes the computer to be used by one terminal at a time. The most commonly employed systems are CP/M, MS-DOS and DR-DOS. PC-DOS is the version of MS-DOS which is used in IBM PC machines, MS-DOS is used in IBM-compatible machines, and DR-DOS is Digital Research's version of DOS, while CP/M, once commonly employed, is now used in only a few computers.

GRAPHICAL USER INTERFACE
DOS and most other operating systems contain such a large number of commands that it is difficult for a user to remember very many of them, and this means that commands are continually having to be looked up in the manual. To simplify the operation of a computer an **operating environment** is often used. This is the name given to software that produces a graphical interface between the user and the computer to shield the user from the complexities of the operating system. Increasingly a **graphical user interface** (GUI) is being employed with computers. In a GUI, on-screen menus and icons are employed to indicate the various functions and tasks that the system can carry out. The cursor is replaced by an arrow whose position on the screen can be moved around under the control of a mouse.

The most popular version of GUI for PCs is MicroSoft's Windows; other GUIs that are presently used include IBM's OS/2 and those that run on the Apple Macintosh. Mainframe computers that employ the UNIX operating system use a GUI that is usually either open-look (used by both AT&T and Sun) or OSF/Motif used by most of the rest of the computer industry. All GUIs include a number of common features:

● Some kind of pointing device such as a mouse.

- Pull-down menus, which are shown across the top of the screen
  — the options each gives are shown below the menu when it is
  pointed to by the mouse-controlled arrow.
- On-screen menus, which can be made to appear or disappear by
  moving the pointer.
- Windows that graphically display what the computer is doing.
- Control mechanisms such as dialogue boxes, buttons and sliders,
  and check boxes that allow the user to turn options on and off.

Most GUIs provide a WYSIWYG feature, i.e. the document that is
printed bears a very close relationship with what is displayed on-
screen. The advantages of a GUI are:

- When the computer is first switched on the user is not faced with
  just a DOS prompt and a flashing cursor.
- The user is not required to have a knowledge of (at least) some
  of the DOS commands or to look up any that are not known before
  an application can be run.
- When an icon is 'clicked' by mouse or by keystrokes, the right
  application is automatically set up.
- The quality of the work done and the productivity of the user
  are both improved.
- A GUI is able to provide an interface that is similar to users of
  different types of computer, e.g. terminals attached to a
  mainframe computer and PCs.

A key difference between PC programs is whether they have been
designed to work under the MS-DOS operating system or with the
MicroSoft Windows GUI. Windows software is easier to use but it
operates more slowly on a given machine than does the equivalent
DOS program. If a PC has Windows installed it is still able to use
DOS programs. DOS programs have been in use for much longer
but in many cases they are being phased out in favour of the Windows
versions. Windows requires a PC with an 80386 processor, or better,
to work properly and quickly.

Multi-user systems

A multi-user operating system allows the facilities of the computer
to be switched from one user to another so rapidly that each user has
the impression that he or she has exclusive access to the machine.
The basic idea of a multi-user system is illustrated in Fig. 4.2. A
number of terminals, which may themselves be PCs, are connected
to the mainframe computer via a **multiplexer**. Terminals that are
situated within the same building as the computer are directly
connected to the multiplexer, but terminals at more distant locations
are connected via a data circuit that consists of a telephone line and
two **modems**. (Multiplexers, modems, and data circuits are considered
in Chapter 11.) Because a considerable part of the mainframe
computer's processing power is occupied with allocating computer

Fig. 4.2 Multi-user computer system.

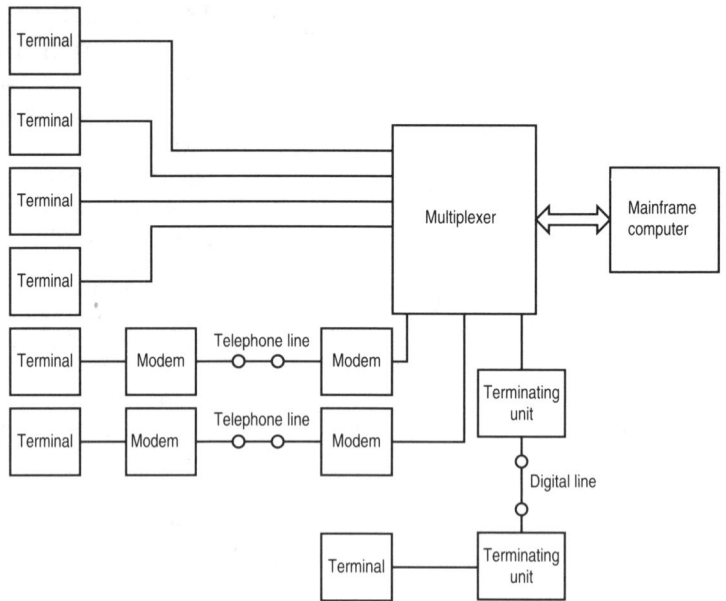

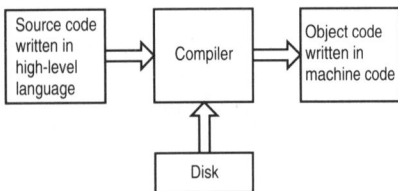

Fig. 4.3 Use of a compiler.

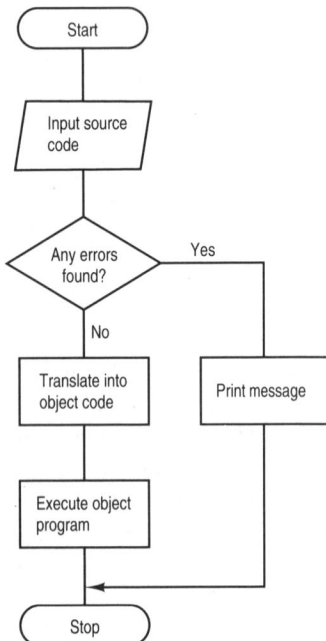

Fig. 4.4 Flow chart of a compiler.

time to the different terminals, the computing power made available to each terminal is often smaller than that provided by a PC. The operating system is complex and the most commonly employed are known as UNIX and VAX/VMS. UNIX is written in a language known as C and is not only multi-user but also multi-task, i.e. a user can perform several tasks simultaneously such as editing one file while printing out another. UNIX has been employed since 1969 for mainframe computers and minicomputers and is now being increasingly employed for PCs, particularly those connected to networks. The PC version is known as UNIXWARE. Some typical examples of multi-user systems are airline booking systems where terminals exist in many travel agent offices, and bank or building society wall cash dispensers.

A better arrangement that allows several users to access a computer is the distributed system; this employs a **local area network** and will be discussed in Chapter 11.

Other examples of systems programs are given below.

COMPILER

A compiler is a complex program which converts a high-level **source** program into machine or **object** code and then the object code is saved. The object program can then be understood by the computer and executed. The compiler translates, or compiles, each high-level language statement (source code) into its machine or object code equivalent. The basic principle of compilation is illustrated in Fig. 4.3. and the steps involved in the translation from source code to object code are shown by the flowchart given in Fig. 4.4. Once the object

Program   Memory

| LET P = A + B |
| LET Q = A ∗ B |

A	1001
B	1002
P	1003
Q	1004
	1005
	1006

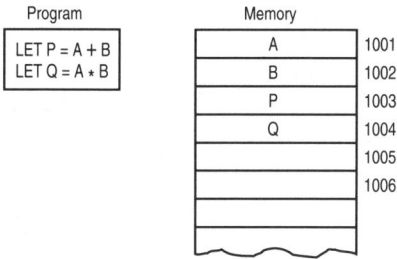

Fig. 4.5 Allocation of memory locations.

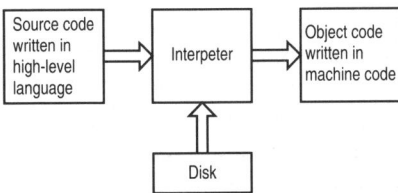

Fig. 4.6 Use of an interpreter.

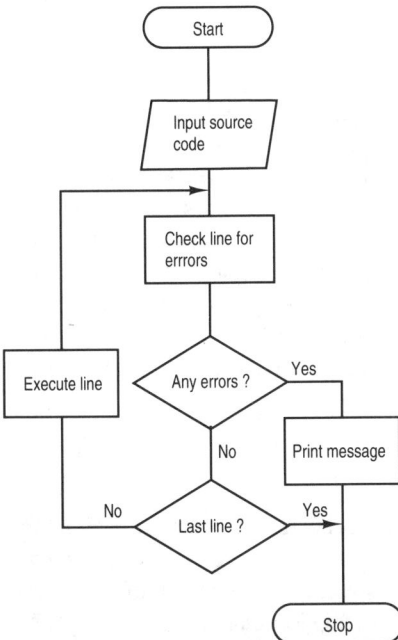

Fig. 4.7 Flow chart of an interpreter.

code has been produced and saved the source code is no longer needed for the running of the program. A compiler generates several machine code statements for each high-level statement and allocates absolute memory location addresses for all the variables and instructions in the high-level program. Thus, in the BASIC statements LET P = A + B and LET Q = A ∗ B, the variables A, B, P and Q will each be allocated a separate location in the main memory, as shown in Fig. 4.5. The compiler also provides error detection and will reject any incorrect statements or logical errors that may exist. Normally, the compiler produces a listing of the source program together with the object code, a list of the memory locations used, and the errors detected. If no errors are found the compiler will instruct the CPU to run the program.

The compiler also incorporates into the object code any library subroutines called for by the programmer and supplies the interconnecting links between the various parts of the program. A library subroutine is a short program that has been written previously and is often used in various programs. The subroutine is held in the backing store and is retrieved whenever it is required for use in a program.

Any particular compiler is able to translate just the one high-level program; if a computer is to use two high-level languages, say BASIC and PASCAL, it will have to be provided with two compilers.

INTERPRETER

An interpreter performs a similar function to a compiler; that is, it also translates a high-level language into the machine code used by a particular computer (see Fig. 4.6). The flowchart showing the steps involved in interpreting a high-level language is shown in Fig. 4.7. An interpreter cannot dispense with the source code; the high-level language — almost certainly BASIC — must remain in memory while the program is running. This is because each statement in the program is first translated and then executed. All the operations used in the language are already compiled and are stored in memory ready for use. In a PC these routines are held in a ROM. The lines of the program are entered into the computer with line numbers to indicate their correct order. Each line in turn is translated into machine code and is then run before the next line is translated. An interpreted language is easier to debug and to edit since any syntax error that may occur in a line is indicated immediately, but it is slower to run than a compiled language. BASIC is an interpreted language and so may be PASCAL. Because of the relative slowness of an interpreter some versions of BASIC are compiled.

ASSEMBLER

An assembler is used to convert a program written in assembly language into the machine code used by the computer. An assembler will also (a) decide where the object program and its data are to be stored, (b) detect errors, and (c) print a listing of the source and object programs.

EDITOR

An editor is used by a programmer as an aid to the development of a program. The editor allows errors made when entering a program to be corrected.

LINKER

A linker program is used to join together a newly written program segment to an existing program.

MONITOR

A monitor program is one which runs automatically immediately the computer is switched on. A monitor is normally resident in a ROM. Its main functions are to allow the user easy access to the computer memory, since the monitor program includes such features as keyboard scanning, output to a display or printer, and conversion from decimal to hexadecimal. Often, the facility is provided to allow the user to enter numbers in hexadecimal form directly into the main memory and to execute programs that are written using that scheme. More sophisticated monitor programs may include such facilities as tracing the execution of a program by allowing direct access to information in a register at each instruction step.

Generally, these systems programs are stored on a disk and are loaded automatically into the computer when it is switched on or is reset.

Support (utility) software

Support programs (also commonly known as **utility** programs) are usually stored on the same disk as the operating system. They are not automatically loaded when the computer is switched on but have to be called by entering the appropriate command at the keyboard. Support programs include a disk formatter, a program to copy files from one disk to another or from tape to disk or vice versa, a sorting program, a merging program, a printer driver, and several more.

Application software

Application software comprises specialized programs that allow the computer to perform a specific task, such as word processing, spreadsheets, databases, communications packages, desktop publishing, and accounts and graphics packages. An application package may be either **command driven** or **menu driven**. Most packages are able to work with different kinds of printer. They include a program called a **printer**

driver which interfaces between the computer and the printer. A printer driver contains the information needed for a computer to work with a particular type of printer. The program details the codes which the printer uses to perform various actions, such as change fonts or print italics. The user has merely to select the name of the printer from an on-screen list of printers handled by the program.

Word processors

A word-processing program makes a computer act as a word processor and it is used to make the production of text as easy as possible. With the program loaded into the computer the user can see on-screen the text as it is typed in at the keyboard. A word-processor package offers many more facilities than are available on an electronic typewriter. For example, any amount of text may be deleted, or added, or moved to another position, and existing text held in a file may be accessed and merged into a new document, a standard letter can be automatically headed with a different name and address for each copy, the spelling can be checked, and so on. The usage of a word processor is particularly advantageous for text that is likely to go through several drafts, like this book for example.

Word-processing programs may be either DOS or Windows based. With a DOS word processor the way in which letters appear on screen is fixed and their style cannot be varied much, if at all. Therefore, the text shown on-screen is similar to that which a typewriter could produce. DOS word-processing programs are more difficult to learn to use than are Windows-based word processors, because the GUI version shows the user what to do while the DOS user has to remember the ways in which the program deals with any particular feature.

Increasingly, word processing is carried out using a Windows environment. The use of a graphics mode, instead of a character mode as with DOS, means that different text sizes, styles, and typefaces can be displayed on-screen. This feature allows WYSIWYG. Thus, a Windows-based word processor is not only easier to use but also has the potential to produce much better results. The on-screen display may be varied as much as necessary until the required document layout is obtained. Windows word processors often offer additional features similar to those given by a DTP package: the import of graphics directly into frames on a page, automatic text wrap-around the frames and easier colour printing.

Databases

A **file** is an organized collection of records and a record is a group of related items of data that may be treated together. A database is a program designed to store, manipulate and retrieve information of all kinds. It allows the user to design and create data files and to retrieve information from those files. Before data can be stored in a database the program must be told which details are to be filed and the type of data each heading represents, e.g. in a list of names and addresses — surname, forename or initials, address, and perhaps telephone number. Each database has its own method by which the user supplies the information and some databases are very much easier to set up than others. An example is given in Chapter 5.

The user of a database is able to store lists of names and addresses, of electrical components and type numbers, etc. The data may be entered into the database in any order and may then be sorted using any required parameter, e.g. names in alphabetical order, or addresses in street order or in town order. A database may be searched for a specified kind of data, such as all those persons with a common surname, or with a common first name. In most databases the data is organized in the form of simple tables: one column might hold names, another column might hold addresses, another column hold telephone numbers, and so on.

The choice of a database package is difficult. The first consideration is the operating system which is used by the computer on which the database is to run. If the computer is a PC then a Windows package could be employed. If the database is to serve multiple users then a network version is called for. Memory requirements can be a problem for DOS users since many database packages need expanded or extended memory to increase their performance. Windows software is not concerned with the difference between conventional and extended memory, so this is not a problem.

The use of a database may require the use of a mouse, particularly if a Windows version is used. Many text-mode DOS products also use a mouse and pull-down menus.

A database may be either **flat-file** (or card-box) or **relational**.

FLAT-FILE DATABASES

A flat-file database operates in a similar way to a card index. It consists of a single data file in which each record has an identical format. A simple example is a file of names and addresses. A flat-file database is unable to pull-in data from other database files so if a user has data items that appear in multiple records then the items must be entered and stored separately for each record.

RELATIONAL DATABASE

A relational database is one that can handle several files at the same time and link together files of related information. The user is able to look up and import information from other files and/or to export information to other files. A relational database is more complex for the person who designs the system but simpler for the person who uses it. This kind of database is needed for more complex applications such as invoicing. Here customer details are taken from one file and product details from another, and the two sets of data are combined.

The customer file would include such details as names and addresses and purchases made, while the product file would contain stock details such as price, stock levels, and stock ordered from suppliers. A relational database allows the user to add, delete, edit, and print out any data entered into the database.

Databases are mainly employed for record keeping where the main requirement is for easy information retrieval. A popular commercial database is dBase, the latest version of which is IV. A large database will probably have a large number of users and hold a large quantity of data. It is necessary to ensure that only one user is able to edit any record at any one time.

Spreadsheets

A spreadsheet is a software package which organizes data in a large grid of columns and rows to give a large number of rectangular boxes or **cells**. Each cell is able to hold data. The basic function of a spreadsheet is to perform specified calculations on data and then, each time new data is entered or the existing data is altered, to immediately and automatically re-do the calculations to give the new correct result. The user must first type in the data and the formulae that specify the calculations to be carried out, and the spreadsheet will then carry out the calculations and display the results. The ease of use and the instantaneous up-dating of results is the main advantage of a spreadsheet.

Although the full grid making up a spreadsheet may have thousands of columns and rows, only a few of the cells are visible on-screen at one time, typically 10 columns and 25 rows. The program allows the user to move a **window** to access any required part of the spreadsheet. The columns in a spreadsheet correspond to the fields in a database and the rows correspond with the records in a database. Each column and row in the grid is given an identifying label. Usually, the columns are given letter labels such as A, B, C, ..., AA, AB, AC, ..., BA, BB, BC, etc., and the rows are numbered. Each cell is then identified by a unique address. Thus, the top left-hand cell is cell A1, the cell just to its right is cell B1, the cell immediately below A1 is labelled A2, and so on. The arrangement is shown in Fig. 4.8.

The user of the spreadsheet can type into each cell either a text item, or a numerical constant, or a formula such as (A1 $\times$ B1) or (A2 $-$ B3) for example. The entered formulae are not visible on-screen when data is entered into the spreadsheet. The range of functions available may include the four arithmetic operators $+$, $-$, $\div$, and $\times$, trigonometric functions such as sin, cos and tan, financial functions such as depreciation and discounted cash flow, and statistical functions such as mean and standard deviation. Each formula expresses a rule by which a cell's value may be calculated. Suppose, for example, that cells A1 and A2 each contain numeric data. If cell A3 contains the formula (A1 $-$ A2) it will display the difference between

	A	B	C	D	E		AA	AB	AC	AD	AE
1	A1								AC1		
2		B2									
3			C3								
4				D4			AA4				
5					E5						

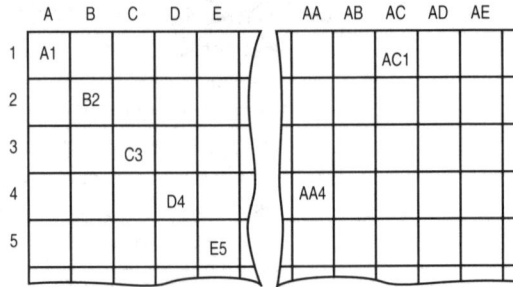

Fig. 4.8 Basic arrangement of a spreadsheet.

the numbers contained in A1 and A2. If the data in either cell A1 or A2 or in both cells is changed, cell A3 will then display their new difference.

Any calculation may use a number taken from another cell, which may, in turn, also be the result of a calculation in yet another cell. A more complex calculation can be broken down into a number of steps and the intermediate results of each step displayed in different cells.

Example 4.1

If cell A1 contains 2000, cell A2 contains 1000 and cell A3 contains the formula (A1 − A2/2), what value will cell A3 display? If the data in cell A2 is changed to 1200 what is then displayed in cell A3?

Solution
A3 displays 2000 − 1000/2 = 1500. (*Ans.*)
A3 now displays 2000 − 1200/2 = 1400. (*Ans.*)

A spreadsheet has a variety of features that make it easy to use. Text can be inserted into a cell to describe the content of other nearby cells. New rows and/or columns may be added if some of the factors in a calculation are altered. The formulae used in conjunction with the spreadsheet, as well as the numbers used, can be stored on disk and the results of a calculation printed out, or displayed on screen — perhaps in chart or graph form — or inserted into another document. In addition to allowing the user to enter and edit data most spreadsheets also possess the ability to remember a sequence of commands that are used to perform a particular task, e.g. to print out a particular section of the spreadsheet. This sequence can then be re-called at any time without the intervention of the user. This kind of predefined sequence of commands is known as a **macro operation**.

The basic applications of a spreadsheet are: doing calculations which involve large numbers of figures that need to be added, subtracted, tabulated, or totalled; carrying out 'What if?' exercises such as 'What would be the total cost of a new car if a down payment of $x\%$ is made with the remaining cost paid off over y years, if the interest rate is

$z\%$?'. Different values of x, y, and z can be entered when the total cost will be instantaneously displayed. The more expensive spreadsheets can be employed for **equation solving**. A spreadsheet is set up to model some required operation. It will include data cells that specify variables in the problem — e.g. batch size, unit costs, and wage rates — and another cell that works out the results. The spreadsheet can then be used to discover the set of inputs that maximize profit. This result may be obtained by altering the constants of the various data cells one by one and observing the resultant profit, or by using a **solver program**. A solver program automatically adjusts the data to obtain the maximum profit and the data cells can then be read to determine the optimum values of the variables. A solver program is very useful for all kinds of problem where some parameters must be maximized, or minimized, or set to some particular values to optimize the results of the operation. Spreadsheets are considered further in Chapter 5.

Desktop publishing

Desktop publishing (DTP) is the use of a computer to emulate the work of both a typesetter and a page make-up artist. A DTP package is used for the preparation of manuals, magazines, and newsletters and offers similar facilities to a word-processing package plus the ability to import images, to produce graphics, to specify the fonts to be used, and to define exactly the layout of each page in a document. A page framework is provided and text is confined to the chosen parts of the frame, leaving space for graphics, images, etc. The word-processing facilities provided may not be comprehensive since often text will have been first created using a word-processing package.

Graphics programs

A number of graphics programs are available that perform vastly differing tasks. Some of these programs are highly specialized while others allow the user to create original artwork. The latter may be either **drawing** or **painting** programs. A drawing program draws lines and shapes to produce a picture and the colour is added by filling in areas that are bounded by the lines. With a painting program a mouse, or the cursor keys, is/are used to move a pointer around the screen leaving a trail of individual blobs of colour wherever required. Both types of program include a facility for text to be placed onto the created images.

Communication software

Communication software is necessary if a computer is to be able to communicate with other computers via a data link. If the data link uses analogue techniques, then peripheral hardware known as a

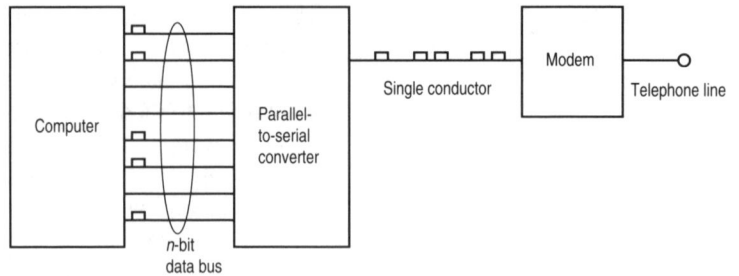

Fig. 4.9 Parallel output of a computer is converted into serial form for transmission over an analogue telephone line.

modem will also be necessary. The communication software, together with the necessary hardware, converts the data outputted by the computer from parallel form into serial form, shown in Fig. 4.9. The function of the modem is to convert the serial bit stream into voice-frequency signals that are able to pass over the public telephone system.

Communication software also enables **electronic mail** to be sent and received and allows a user to access **bulletin boards** and distant databases.

Integrated packages

An integrated software package incorporates several different application programs. Typically, an integrated package might include a word processor, a spreadsheet, and a database. To some extent it is generally possible to move data and/or results between the individual software modules. All the programs will use the same menu structure and have the same appearance, which means that there is less learning demanded of the user and data can be transferred from one module to another. An integrated package covers several purposes but none of them in any great depth.

Modern software packages with GUI (like Windows) allow the user to tailor the operation of the program to individual requirement. A database manager may include data entry forms and reports on-screen and produce complex applications without having to do anything more complex than choosing commands from a menu. A word processor may include a macro language that helps the user automate frequent tasks with the minimum of effort and skill.

The aids provided with a GUI-based program to permit modifications to be made are such that the program waits for the user to do something and then reacts accordingly. The program can perform a wide range of routines and the user has only to select the routines that are to be run, the data the routines are to be given, and how the results are to be presented on-screen.

Flow charts and pseudocodes

Before a program is written the task to be performed by the computer needs to be clearly defined. An algorithm defines a sequence of steps that a computer will be able to follow to solve a problem. It is quite possible that a given problem has more than one method of solution and then more than one algorithm can be obtained. Aids to this process are the **flow chart** and **pseudocode**.

Flow charts

A flow chart is a diagrammatic representation of a logic process that graphically shows an algorithm and indicates a method for its solution. The chart shows the **flow** (or order) of the required program. A flow chart can be used to show how a particular problem could be solved, to aid program development and debugging and the checking for logical errors, and to assist in program documentation. Each step in the algorithm is represented by a standard symbol in the flow chart. The standard symbols which are used are shown in Fig. 4.10. The *terminator* (Fig. 4.10(a)) shows the symbol for the start and end (stop) of a chart. The *processor* (Fig. 4.10(b)) represents a process or action taken that occurs in the algorithm. The process can include any simple operation that causes a change in the value, or in the location, of some data. The *decision maker* (Fig. 4.10(c)) indicates a point at which a decision must be made. A decision box is normally entered from the top and it represents a point in the flow chart at which different processing paths may be chosen. A decision box must have at least two outputs and the five main possibilities can be seen in Fig. 4.11. The *input* or *output* (Fig. 4.10(d)) relates to any operation that inputs or outputs data.

Messages are placed into each symbol to indicate briefly what its function is. Flow charts were commonly employed in the past but

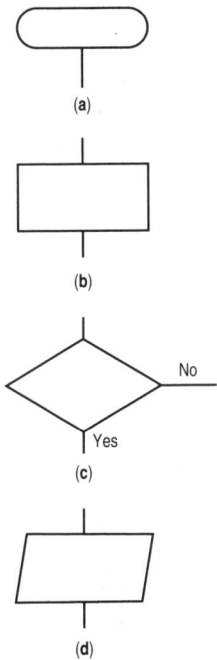

Fig. 4.10 Standard flow chart symbols.

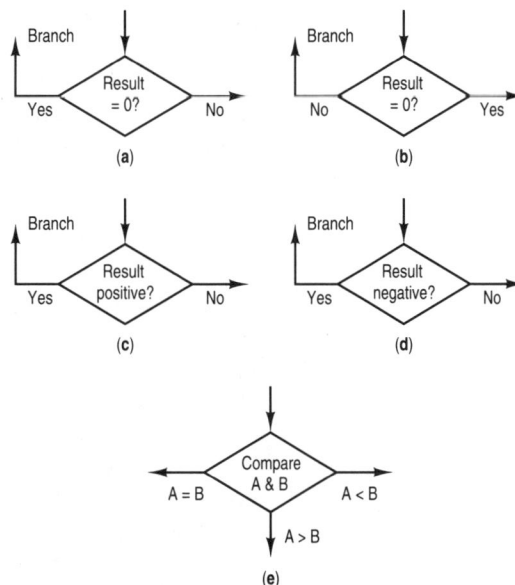

Fig. 4.11 Decisions in a flow chart.

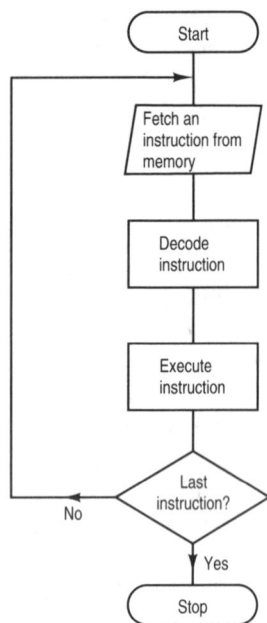

Fig. 4.12 Fetch–execute cycle.

now alternative methods, one of which is known as **pseudocode**, are often used.

Drawing a flow chart is often the first step taken in the development of a computer program. A programmer should plan a program in a logical manner and the flow chart provides one method by which this aim may be achieved. The use of a flow chart is particularly important for a large program since different parts of the program may be interconnected in various complex ways. Once a flow chart has been drawn it makes it easier for other people to understand how the program works. Also a flow chart can be translated into whatever high-level language is to be employed. Several simple examples of flow charts are given below.

The fetch/execute cycle of a computer may be represented by the flow chart shown in Fig. 4.12.

A simple flow chart to solve the equation

$$[(AB + C) - B]/B$$

is shown in Fig. 4.13, where A is a number stored in a memory location and B and C are fixed numbers. The answer is to be stored in a memory location that has been labelled as ANSWER.

Two of the basic circuits used in digital electronics are the AND gate and the OR gate. The AND gate has an output at logic 1 only if all of its inputs are also at the logic 1 level. An OR gate has an output at logic 1 if any one, or more, of its inputs are at logic 1. The flow charts for (a) an AND gate and (b) an OR gate are given in Fig. 4.14.

A loop is a part of a program that may need to be repeated a number of times. There are three kinds of loop structure:

- DO WHILE: The flow chart for a DO WHILE loop is shown in Fig. 4.15. The operations specified by the boxes marked A and B are repeatedly carried out while a particular condition exists; once that condition ceases to exist the loop terminates. If the condition does not exist initially the operations A and B are never executed. The condition could be that a variable is greater than, or less than, or equal to, some fixed number or to another variable.
- REPEAT UNTIL: The REPEAT UNTIL or DO UNTIL loop action is illustrated by the flow chart in Fig. 4.16. The operations A and B are repeatedly carried out until the specified condition is satisfied, when the loop comes to an end. The two operations will always be carried out at least once even if the specified condition never exists.
- FOR NEXT: The flow chart for a FOR NEXT loop is given in Fig. 4.17. A variable COUNT is set to an initial value and its value is tested in the loop. If its value is not equal to a specified value the operations A and B are carried out and COUNT is either incremented or decremented. When COUNT reaches its final

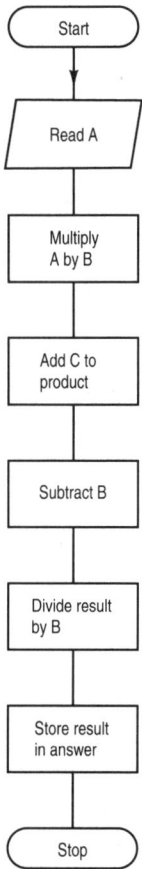

Fig. 4.13 Flow chart for
[(AB + C) − B]/B.

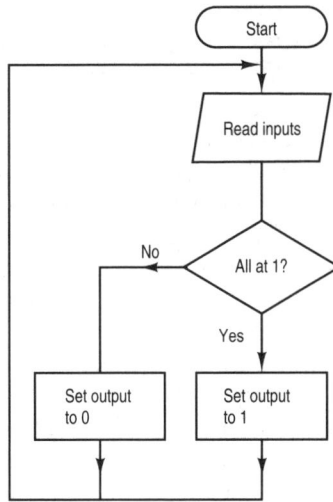

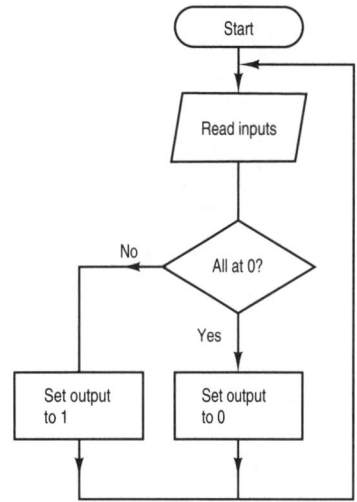

(a)　　　　　　　　(b)

Fig. 4.14 Flow chart for (a) an AND gate and (b) an OR gate.

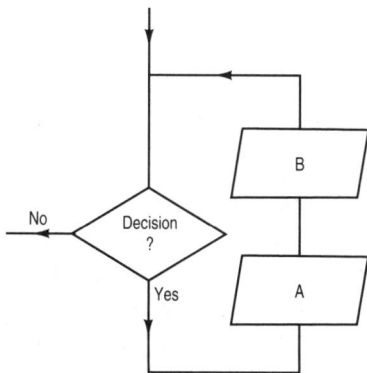

Fig. 4.15 DO WHILE loop.

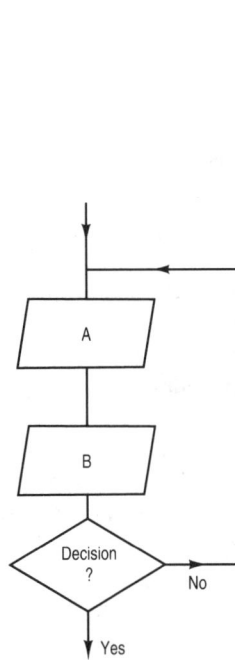

Fig. 4.16 REPEAT (DO) UNTIL loop.

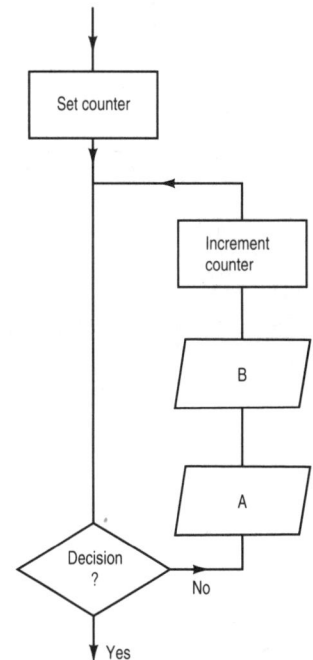

Fig. 4.17 FOR NEXT loop.

value the loop is terminated. Most often the loop counter consists
of a register that initially contains the number of loops to be
executed. On each pass of the loop the number is decremented
and when it reaches 0 the loop is terminated.

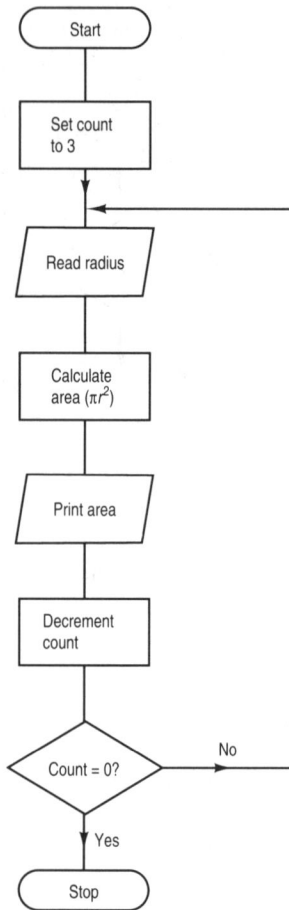

Fig. 4.18 Areas of three circles.

The use of a loop in a simple flow chart is shown in Fig. 4.18. The flow chart represents the problem of calculating the areas of three circles when, in turn, their radii are entered into the computer. The loop is set up by setting COUNT to 3 initially and then decrementing COUNT once in each pass through the loop. The loop is repeated until the end condition COUNT = 0 is satisfied.

A further illustration of the use of a loop in a flow chart is shown in Fig. 4.19; this chart is for the algorithm that gives the solution to the problem of adding up all the integer numbers from 1 to 20 and storing the result of the addition in a memory location labelled RESULT. The sum of the addition is labelled SUM and the latest number to be added to SUM is labelled VALUE. The loop is used to repeat the addition of the current SUM and the latest value of the number to be added. In this loop the initial value of the loop counter VALUE is set to zero and the count is incremented in each pass over the loop.

There is no limitation to the number of loops that may be used in a flow diagram and a more complex example is shown in Fig. 4.20. A computer controlled process is used to sort packages into one of three weight categories: A, weight $W < 1$ kg; B, weight $W > 2$ kg; C, weight $W \geq 1$ kg and ≤ 2 kg. A count is kept of the number of packages, N_1, N_2, and N_3, in each category and also of the total number T of packages. The process is to stop if the number of packages in either category A or category B exceeds 5 or when category B = 100.

As another example of the use of multiple loops Fig. 4.21 shows the flow diagram of an electronic lock that contains 5 loops.

The process of converting the algorithm represented by a flow chart into a program is known as **coding**.

Pseudocodes

The use of a pseudocode is an alternative method to flow-charting for the planning of a program which provides a written description of how a problem could be solved. A pseudocode looks rather like a program but it is not tied to any particular language. The necessary steps in the program are shown in shortened form in a way that makes the whole program flow clear. Another advantage of the method is that there is no particular form of pseudocode laid down so that each programmer can use whatever version best suits the purpose. A pseudocode is easier to write and to follow than a corresponding flow chart. If the language to be used for the program has already been decided upon, the pseudocode may as well be written on the same lines. Pseudocode versions of the flow charts given in Figs 4.18 and 4.19 are as follows:

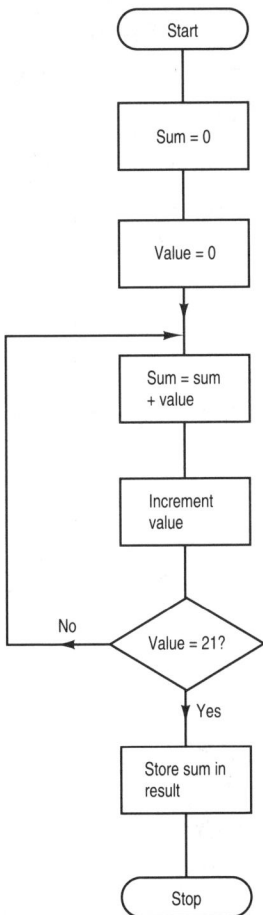

Fig. 4.19 Sum of all integer numbers from 1 to 20.

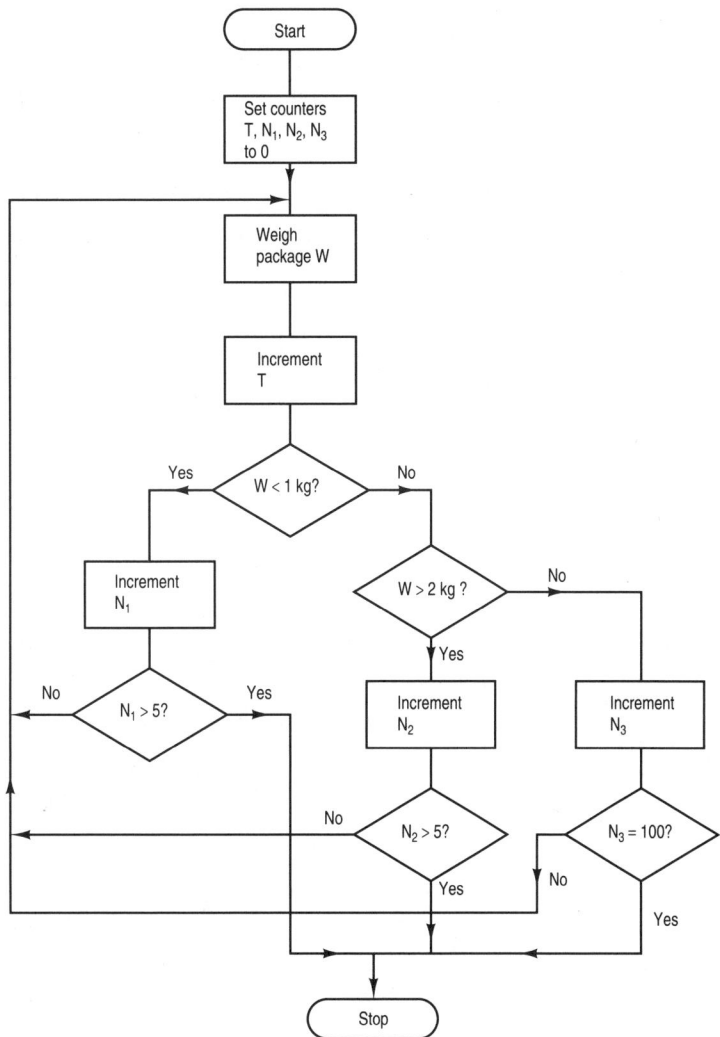

Fig. 4.20 Flow chart with several loops.

- Radii of three circles

  Count := 3
  REPEAT
    READ radius r
    area = πr^2
    PRINT area
    DECREMENT count
  UNTIL count = 0

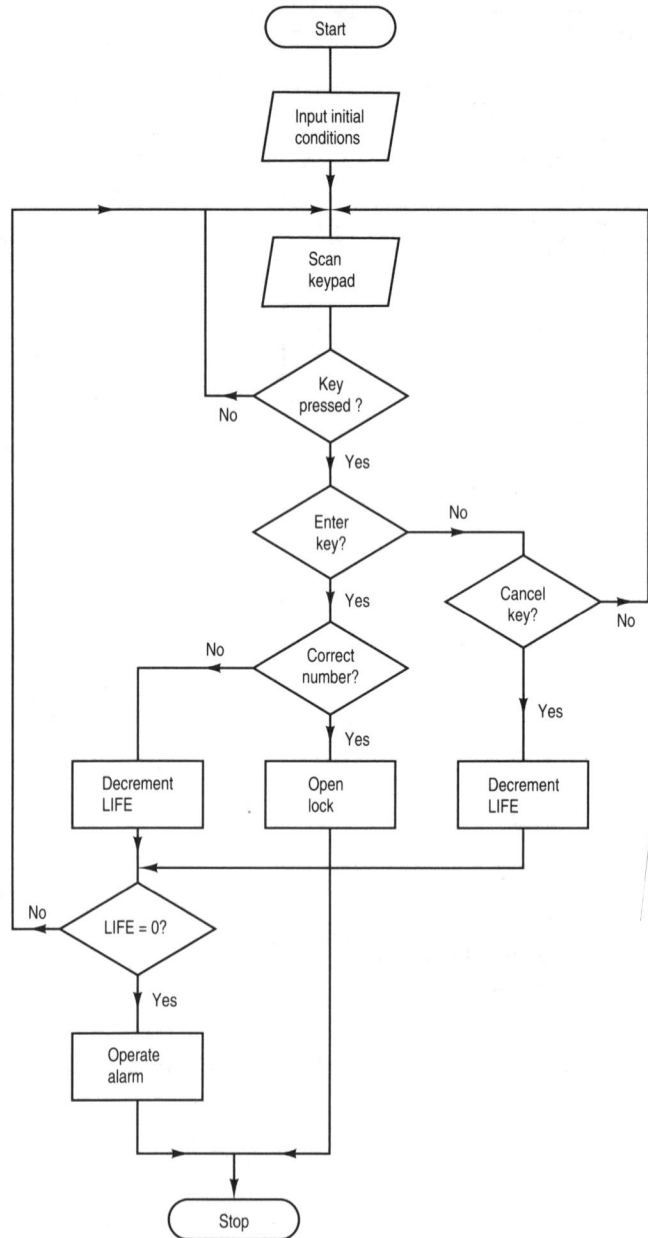

Fig. 4.21 Electronic lock.

- Total of numbers 1 through to 20

    Sum := 0
    Value := 0
    REPEAT
      Sum = Value + Sum
      Value = Value + 1
    UNTIL Value = 21

Simple programming

A program consists of a sequence of instructions. A complete instruction plus data is known as a **statement** and most languages require that there is one statement in each line of the program. A **command** is an outside request for the computer to do something, e.g. RUN or LIST the program.

When writing a program the programmer must use only allowable mnemonics or reserved words since any other entries will be neither recognized nor acted upon by the assembler/compiler/interpreter.

Low-level languages

An assembly language can be very efficient and quick to operate, but it is difficult both to write and to understand once written. An assembly language is seldom used for application programs, but is often used for systems and support programs. C is a low-level language containing instructions that are similar to those used in machine code. This means that C is also a very efficient language and it is used to write some operating systems.

An **instruction set** is the set of machine code instructions that a computer is able to understand. The instruction sets of the processors used in mainframe computers and in minicomputers contain a large number of instructions while the sets for PCs may be less comprehensive. Instruction sets vary from one computer to another but a basic understanding of the writing of assembly language programs can be gained by writing programs using a simple instruction set containing the following instructions:

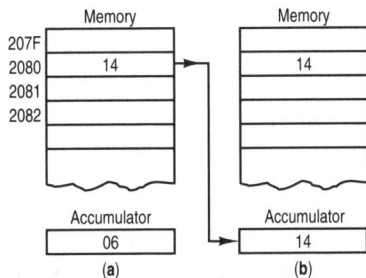

Fig. 4.22 Load accumulator.

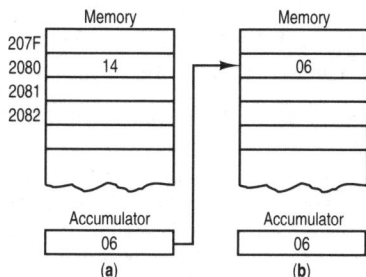

Fig. 4.23 Store contents of accumulator.

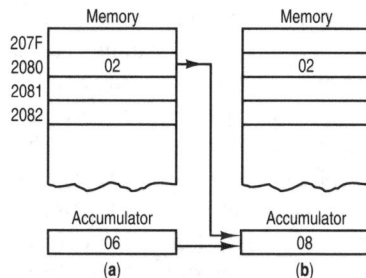

Fig. 4.24 Add two numbers.

- LOAD: this instruction means load the accumulator. Figure 4.22(a) shows the contents of the accumulator as 06, and the contents of memory location 2080 as 14 before the instruction is executed. After the LOAD instruction has been executed (Fig. 4.22(b)), 14 has been loaded into the accumulator and the original 06 lost. The contents of location 2080 have not been changed.
- STORE: store the contents of the accumulator. Figure 4.23 shows an example of what happens when the STORE instruction is executed. As in the previous figure, before execution the accumulator contains 06 and the memory location contains 14. After the STORE instruction has been executed the contents of the accumulator are unchanged but the contents of the memory location have been altered to become 06.
- ADD: add a number or the contents of a memory location to the contents of the accumulator. Figure 4.24(a) shows the contents of the accumulator and a memory location, before the ADD instruction is executed, to be 06 and 02 respectively. Figure 4.24(b) shows the contents of the accumulator and the same memory location after the instruction has been executed.
- DECREMENT: reduce the contents of a specified register or

memory location by 1. The number in the memory location, or the register, is decremented by moving it to the accumulator, reducing its value by 1, and then moving it back to its original memory location, or register.

- INCREMENT: increase the contents of a specified register by 1. The INCREMENT instruction is carried out in similar manner to DECREMENT.
- HALT: stop the program.
- CONDITIONAL JUMP: a conditional jump, or branch, instruction is executed only if a specified condition is satisfied. Some examples include: BNE, branch if result is not equal to 0; BEQ, branch if result is equal to 0; BPL, branch if result is plus; BMI, branch if result is minus. When such an instruction is reached the CPU will test to see if the specified condition exists. If it does, the branch will be made; if it does not, the program will continue on to the next instruction.
- UNCONDITIONAL JUMP: a JUMP instruction causes the CPU to jump from its current position in the program to some other position. Examples are JSR, jump to subroutine, and JUMP (or BRA), jump always to a specified memory location.
- COMPARE: compare the contents of a specified memory location, or of a specified register, with the contents of the accumulator. The contents of the memory location are subtracted from the contents of the accumulator and signalled as being (a) equal, i.e. the difference is zero, (b) greater than, i.e. the difference is negative, or (c) less than, i.e. the difference is positive. The contents of both the accumulator and the memory location, or register, are not changed by the comparison.

Most of the instructions obeyed by a computer can be placed into one of three main categories:

1. **Data transfer instructions**. These are used to transfer data between different registers in the CPU, between a register and the ALU, or between a register and the main memory. Some examples of such instructions are: LOAD ACCUMULATOR; COPY A,B (copy the contents of register A into register B), and STORE (store the contents of the accumulator in a specified memory location).

2. **Arithmetic and logic instructions**. The data manipulation instructions are used to perform arithmetic operations such as add and subtract, and logic instruction like AND and OR. The increment and decrement instructions are also included. Multiply and divide instructions are provided on the larger computers but not on all PCs. Some examples follow: ADD X (add the contents of register X to the contents of the accumulator); SUB Y (subtract the contents of register Y from the contents of the accumulator); INCREMENT and DECREMENT (add 1 to, or subtract 1 from, the contents of a specified register).

3. **Test and branch instructions**. Test and branch (also known as

flow and control) instructions are used to transfer program control to another part of the memory. The instruction tests (checks) various data values, such as whether or not a value is zero, or positive, or negative and, according to the result of the test, transfers control to another part of the program. It is this kind of instruction that gives rise to a departure from the normal sequence of program steps known variously as jumps, branches and loops. This is known as **conditional branching or jumping**. Some examples have been given previously.

Symbolic addresses are given to memory locations so that the assembler is able to identify them. This ensures that the programmer does not need to know where each location used by a program is in the memory. Further, if a program is altered in some way parts of it may well be moved in the memory, but although the absolute address might have changed the symbolic address will remain the same. The accumulator may be loaded with a number, often indicated by the symbol #. Thus LOAD #23 means load the accumulator with decimal number 23, this is known as **immediate addressing**. Alternatively, the accumulator may be loaded with the contents of a particular memory location. This location may be given a symbolic address. Thus the instructions

    LOAD A #23
    STORE A FIG

mean load the accumulator with number 23 and then store that number in the memory location whose symbolic address is FIG.

It is necessary to have first told the assembler where the memory location FIG is, and there are two ways in which this may be done:

- Start the program with a directive that specifies the location of FIG; thus, if this location is chosen by the programmer to be hex 00FA then the first line of the program should be

    FIG = $00FA

  [The symbol $ is often used to indicate a hex number and the % symbol to indicate a binary number.]
- Other processors require the directive to be entered in a different way, and

    FIG EQU $00FA
  or
    FIG EQU 00FAH

  are two commonly employed methods. EQU is a **directive** that is used to assign a value to a label.

Hence the program segment

    FIG = $00FA   FIG EQU $00FA
    LOAD A #23   or   LOAD A #23
    STORE A FIG   STORE A FIG

will store the number 23 in the memory location 00FA.

The programmer must be very careful with his choice of memory locations to ensure that locations reserved for other purposes by the computer or for other programs are not used. It is also necessary to tell the assembler where the machine code, or object code, program is to be stored and the location for the first instruction should be given. This is done in different ways for different CPUs and another often used directive is ORG. ORG is used to define the starting address of a program. Directives are used to name the program, to define its starting address, and to assign symbolic labels to memory locations.

Assuming that the start address for the program is to be 0C00H, this is indicated by

ORG $0C00   or   ORG 0C00H

The simple program has now become:

ORG $0C00	or	ORG 0C00H
FIG = $00FA		FIG EQU 00FAH
LOAD A #23		LOAD A #23
STORE A FIG		STORE A FIG

If the data to be manipulated by the computer is obtained from the outside world via the input/output circuitry, then, depending upon the processor employed, either the input/output port is treated like a memory location or INPUT/OUTPUT instructions are provided.

All computations and comparisons are carried out by the ALU. For a calculation or a comparison instruction to be executed the numbers to be used must be taken from memory and placed into the ALU. If, for example, two numbers are to be added together the necessary instructions are:

- Take one of the numbers from memory and place it into the accumulator. The instruction is LOAD A followed by the address of the memory location, or the register, where the first number is stored. If symbolic addressing is used and the location label is NUM 1 then the instruction is LOAD A, NUM 1. If the number is held in a register, say register X, then the instruction would be LOAD A, X.
- Add the second number to the number now held in the accumulator. The second number may be stored either at a memory location or in a register. If it is stored at location NUM 2 then the instruction is ADD, NUM 2, but if it is stored in, say, register Y then the instruction is ADD, Y.
- Store the result of the calculation in memory location SUM. The necessary instruction is STORE A, SUM.

Thus the simple program could consist of any combination of the above. Different microprocessors would write each of these instructions in a slightly different way; examples are given in Chapter 8.

Two-pass assembler

Many assemblers require that the source code is assembled twice. If, on the first **pass**, the assembler meets a branch instruction that calls for a forward move to an as yet unassigned symbolic address it will treat it as an undefined label error. This means that the symbolic address will not yet be assigned a memory location. As the assembler continues its first pass through the source code it will eventually meet the symbolic address called and can then assign locations to it. On a second pass through the source code the location of each symbolic address met is already known and the correct location can be assigned to produce the final object code.

The main features of an assembler are as follows:

- The programmer is able to use symbolic addresses. This means that the assembler will work out the absolute memory addresses for all labels used in a program.
- Any errors found during the assembly of a program will be signalled.
- It can produce a listing of the source code and/or the object code, together with any error messages.
- It will work out all necessary forward and backward references, such as branch instructions and subroutines.

Loader

A loader is a program whose function is to load the object code produced by an assembler into the memory of the computer.

Example 4.2

Draw the flow chart of, and write, a program to add the contents of memory address NUM 1 to the contents of address NUM 2 and to then store the result of the addition at memory address SUM. Start the program at memory location 0200H and choose locations for NUM 1, NUM 2, and SUM.

Solution
The flow chart is shown in Fig. 4.25.

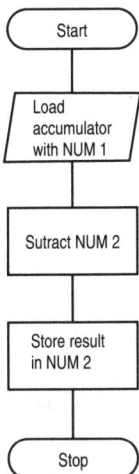

Fig. 4.25 Example 4.2: flow chart.

```
ORG $0020
NUM 1 = $0300
NUM 2 = $0302
LOAD A NUM 1
ADD NUM 2
STORE A SUM
HALT
```

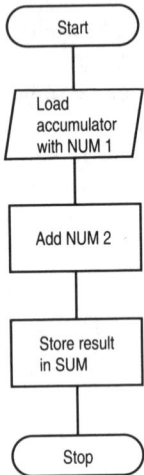

Fig. 4.26 Example 4.3: flow chart.

Example 4.3

Write a program to subtract the number stored in memory location 0050H from the number stored in location 0060H. Store the result at location 0060H. Start the program from memory location 0020H. Draw the flow chart.

Solution

The flow chart is shown in Fig. 4.26.

```
       ORG $0020
NUM 1  EQU $0050
NUM 2  EQU $0060
       LOAD A NUM 1
       SUBTRACT NUM 2
       STORE A NUM 2
       HALT
```

The equivalent to a FOR NEXT loop in a flow chart can be obtained in two ways using a register, say X, as a counter.

(a) FOR A = 1 to 20; NEXT:

```
          LOAD X # 1
LOOP      Start of the program segment
          "   "   "     "        "
          "   "   "     "        "

          End of program segment
          INCREMENT X
          COMPARE X, 21
          BNE LOOP       /If not zero branch to symbolic
                          address LOOP
```

(b) FOR A = 1 to 20 STEP -1; NEXT:

```
          LOAD X, 20
LOOP      Program segment
          DECREMENT X
          BNE LOOP
```

Example 4.4

Write a program to delay a computer for a short time.

Solution

```
        ORG $0200
        DELAY = $0300
        LOAD A # 1000
DELAY   DECREMENT A
        BNE DELAY         /If the contents of the accumulator are
                           not zero branch to symbolic address
                           DELAY

        HALT
```

[Note that the delay obtained depends upon the value loaded into the accumulator.]

Example 4.5

Write a program to store the sum of the numbers 1 through to 20 in a memory location.

Solution

```
        ORG $0200
        SUM EQU $0300
        LOAD A #00
        STORE SUM #00        /Ensure SUM is empty
        LOAD X # 20
LOOP    LOAD A, X
        ADD SUM
        STORE A SUM
        DECREMENT X
        BNE LOOP             /If the contents of register X are not
                             zero branch to symbolic address LOOP
        HALT
```

Alternatively,

```
        LOAD X, #00          /Clear register X
LOOP    LOAD A, X
        ADD SUM
        STORE A SUM
        INCREMENT X
        COMPARE X, 21
        BNE LOOP             /If the contents of the X register are not
                             21 break to symbolic address LOOP
        HALT
```

Example 4.6

Write a program to store the numbers 0, 1, 2, through to 9 in memory locations $0420 onwards. Start the program in location $0200.

Solution

```
        ORG $0020
        LOAD X #00
        LOAD A #00
START   STORE A, $(0420 + X) /Store the contents of the accumulator
                             in memory location (0420 plus contents
                             of X register)
        INCREMENT X
        LOAD A, X
        COMPARE X, 10
        BNE START
        HALT
```

Alternatively,

```
        ORG $0200
        LOAD X #10
        LOAD A #00
START   STORE A, $(0420 + X)
        DECREMENT X
        BNE START
        HALT
```

Example 4.7

Write a program to implement the pseudocode shown.

```
BEGIN
    SUM = X + Y
    IF SUM ≤ 20
    THEN SUM = SUM + 5
    ELSE SUM = SUM + 10
    END IF
END
```

Solution

```
            ORG $0200
            X EQU $0260
            Y EQU $0261
            SUM EQU $0262
            LOAD A, X
            ADD Y
            STORE A, SUM
            COMPARE SUM, 20
            BMI LESS
            LOAD A #5
            ADD SUM
            STORE A SUM
            JUMP STOP
LESS        LOAD A #10
            ADD SUM
            STORE A SUM
STOP        HALT
```

High-level languages

A high-level program may be **structured**, or it may be **non-structured**. A structured program is one that is divided into a series of self-contained units, which may be either **procedures** or programmer-defined **functions**. Each procedure or function generally performs a particular task, e.g. a time delay. This task is performed whenever the procedure, or function, is called; when it is completed execution is returned to the line in the main program immediately after the line in which the procedure, or function, was called. The aim of structured programming is to reduce a complex task into a set of smaller and simpler tasks. A structured program is easier to follow, and to understand, than a non-structured program in which execution jumps about the program following JUMP and GOTO instructions.

Subroutines

Very often in a program the same set of instructions is required again and again. For example, an accounting software package might require

```
1000 ┌──────────────────┐
     │                  │
     │   Main program   │
     │                  │
1500 ├──────────────────┤
     │    CALL 5000     │──────┐
1501 ├──────────────────┤      │
     │                  │      │
     │                  │      │
     │   Main program   │      │
     │                  │      │
     │                  │      │
5000 ├──────────────────┤      │
     │    Subroutine    │◄─────┘
5020 ├──────────────────┤
     │     RETURN       │
     └──────────────────┘
```

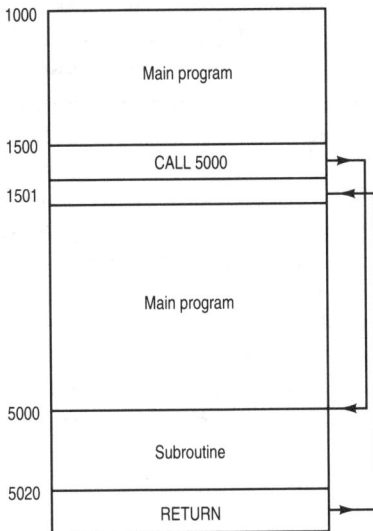

Fig. 4.27 A subroutine.

VAT calculations to be made at various points in the main program. To avoid the need to write these instructions several times within the same program they can be placed within a **subroutine** and called whenever wanted. A subroutine is a block of program within a larger program which performs one specific task. The main program is executed line-by-line until the subroutine is called. The CPU then goes to the specified part of the program and continues executing the instructions in the subroutine until it reaches the instruction RETURN at the end of the subroutine; the CPU then returns to the line in the program that immediately follows the line in which the subroutine was called. An area of the main memory, called the **stack**, is used to store the RETURN addresses so that the CPU knows where to go when it exits the subroutine. When the instruction RETURN is executed the contents of the stack are placed into the program counter.

The method of calling a subroutine varies with different processors but two methods employed are CALL and GOSUB.

The principle of a subroutine is shown in Fig. 4.27.

Procedures

Procedures are employed in structured programming languages, including BBC BASIC. Essentially, a procedure is a kind of subroutine that is called by name rather than by line number. This allows a procedure to be found by the CPU far more quickly than it can find a subroutine. This is because the CPU knows where a procedure is located and so it does not need to search right through the program to find it, as it does to find a subroutine. Also, since each procedure has a different name it makes it easier to follow and understand the program and each procedure can be individually written and debugged.

Languages

The first high-level language, known as FORTRAN, was designed for scientific and engineering calculations and it is very good for number processing. FORTRAN is difficult to read and it is poor at handling strings. Other early high-level languages are ALGOL and COBOL; the former is also an engineering/scientific application orientated language while the latter is business orientated. Vast numbers of commercial and business programs have been written in COBOL and it is still commonly used today many years after its first appearance. COBOL was designed to handle large amounts of data very efficiently and COBOL programs are both easy to read and very good at file handling. The language is particularly suited for commercial applications such as stock control, sales ledgers, and wages/salaries calculations. In COBOL all statements look like sentences in English. LISP is a language designed specifically for use with artificial intelligence research and development; it is available in both compiled and interpreted versions. BASIC was first introduced in 1964 and has since gone through several stages of development until today there are several versions in use, some structured and some not. In most versions of BASIC every line in a program must be numbered but a line may contain more than one statement. Both interpreted and compiled versions of BASIC exist. BASIC is easy to learn and to read but it is slow compared with most other languages. FORTH was originally developed as a language for the computer control of radio telescopes but now it has many other applications in the control field. LOGO is primarily a graphics language that was developed from LISP and it is often used to introduce children to the basic ideas of computer programming. PASCAL is a powerful, yet compact, structured language that is often used for engineering and scientific applications. PASCAL has good file and number handling features but it is not easy to learn. COMOL is a language that combines the best features of BASIC and PASCAL. C is a compiled language that is used for the UNIX operating system and has a similar structure to PASCAL; its main advantage is that it is economical with its memory requirements but it is difficult to learn and to read. PROLOG has some similarities with LISP; it is easy to read but difficult to learn and is not very good at handling numbers.

The vocabulary of a high-level language consists of a number of reserved words that must not be otherwise employed. In BASIC, for example, reserved words include LET, PRINT, INPUT, and READ, while COBOL reserved words include ACCEPT, LABEL, REPORT, and TABLE. A reserved word informs the interpreter/compiler about a particular instruction that the programmer wishes to use in a program. A high-level language requires fewer instructions to perform a given task than does a low-level language. Consider, for example, the BASIC instruction

LET RES = 17.

In assembly code this would become

```
ORG $0200    /Tell the assembler where the first instruction can
             be found
RES = $0220  /Tell the assembler where the location RES is
LOAD £17     /Load the accumulator with 17
STORE RES    /Store contents of accumulator in location RES.
```

BASIC

BASIC is generally thought to be the easiest language to learn and this has led to its being commonly employed for applications that use a small computer. There are several different versions of BASIC, some of which are structured while others are not, but the principal statements are the same in all. In most versions of BASIC the lines in a program are numbered. The following list gives some of the principal statements used in BASIC:

- INPUT: allows data to be entered at the keyboard.
- PRINT: displays data on screen.
- LET: sets a numerical value to a variable.
- GOTO: transfers control to another line in the program.
- IF THEN: obeys a command only IF a specified condition is satisfied.
- FOR NEXT: obeys a section of program a specified number of times (giving a loop).
- GOSUB RETURN: enter, and then leave, a subroutine.
- PROC(A): go to the procedure labelled as A.

Commands used in BASIC include:

- RUN: tell the computer to execute the program.
- LIST: displays on-screen the program held in memory.
- NEW: clears any program stored in memory.

Example 4.8

Write a BASIC program to calculate the area of a circle.

Solution
```
10 area = 0
20 INPUT "Radius of circle?"; R
30 area = 3.142 * R * R
40 PRINT area
50 END
```

Example 4.9

Write a BASIC program to calculate the average of several numbers.

Solution
```
10 INPUT "How many numbers"; N
20 SUM = 0: AVERAGE = 0
30 FOR A = 1 TO N
40 INPUT "Number"; X
50 SUM = SUM + X
60 NEXT A
70 AVERAGE = SUM/N
80 PRINT "Average is"; AVERAGE
90 END
```

Structured versions of BASIC include BBC BASIC, HP BASIC, True BASIC, and Q BASIC. Visual BASIC is a language produced by MicroSoft for writing applications for Windows.

PASCAL

A PASCAL program has an orderly block structure and is usually written by first specifying its overall function and then breaking it down into separate blocks, i.e. a PASCAL program is structured. Either lower-case or upper-case letters may be used at any time since the language does not distinguish between them. All PASCAL programs have three main parts:

1. **Heading**. A heading which consists of a single line of the format

   PROGRAM Name (INPUT, OUTPUT);

   The INPUT, OUTPUT in the brackets indicates to the computer that input data is to be expected and the results are to be outputted.
2. **Declaration**. All constants and variables must be declared before the main body of the program and must be specified as being either REAL or INTEGER numbers.
3. **Statements**. This part lists the actions that are to be carried out. The main body of the program is placed between BEGIN and END statements. The END statement at the final part of the program must be followed by a full stop so that the computer knows when the program has finished. All statements must terminate in a semi-colon. Lastly := means 'takes the value of', or 'becomes equal to'.

Example 4.10

Write a program in PASCAL to calculate the area of a circle.

Solution
```
PROGRAM Circle (INPUT, OUTPUT);
CONST
  pi = 3.142;
VAR
  radius, area: REAL;
BEGIN
  WRITE ('Enter radius')          /Causes 'Enter radius' to appear on-
                                   screen
  READ (radius)                   /Causes keyboard entry to be entered
                                   into the computer
area: = pi * radius * radius;
  WRITELN ('area is', area, 'cm²')/As for WRITE but makes a new line
                                   follow
END.
```

Example 4.11

Write a PASCAL program to calculate the average of 6 numbers.

Solution
PROGRAM Average (INPUT, OUTPUT);
VAR
  count: INTEGER;
  number, total, average: REAL;

Procedure numbers

  BEGIN
    FOR count: = 1 TO 6 DO
      BEGIN
        READ (number);
        total: = total + number;
        average: = total/6;
    END;
BEGIN
  WRITELN ('Enter 6 numbers');
  numbers;
    WRITELN ('average is', average);
END

COBOL

Computations in COBOL may be expressed either in mathematical form or by using English words. This is shown by the following example which is carried out using both methods.

Example 4.12

Write a program in COBOL to calculate the area of a circle.

Solution
COMPUTE AREA: = 3.142 * RADIUS * * 2
        [* * means raised to the power of]
or
MULTIPLY 3.142 BY RADIUS BY RADIUS.

Example 4.13

Write a COBOL program to calculate the average of several numbers.

Solution
COMPUTE AVERAGE: = (A + B + C + D + E + F)/6
or ADD A TO B TO C TO D AND DIVIDE SUM BY 4.

Artificial intelligence

Artifical intelligence (AI) is the ability of a computer system to perform a task which, if performed by a human being, would be regarded as intelligent. Examples of this which clearly illustrate the concept are the use of a computer to play chess well and 'understanding' the spoken or written word. There are four main areas of AI: namely, **expert systems, natural languages, vision systems**, and **robotics**.

Expert systems

An **expert system** has been designed to solve problems in a particular field. The system is provided with knowledge of that field by human experts and is told how this knowledge is to be interpreted and used. Many different types of experts system are in existence, such as:

1. **Medical**. The symptoms of a large number of different diseases and illnesses are put into a database along with a number of diagnoses that have been made in the past and found to be correct. The data is derived from a number of medical experts. A doctor is able to enter a patient's symptoms into the system and read out a diagnosis of the problem. The diagnosis is obtained much more quickly than by any alternative method and is more likely to be correct. One medical expert system is used to produce a diagnosis from the results of blood tests.
2. **Credit authorization**. A database can be used to hold information about a person's credit history such as loans that have been taken out and his or her record in making repayments. When further credit is applied for, the person's details and the amount required to be borrowed can be entered into the system which will make a decision about whether or not the request should be granted.
3. **Insurance**. An expert system may be used by a financial adviser to tailor a suitable package for a client. Details of the client's requirements can be entered into the system and some alternative packages outputted.
4. **Training**. Many organizations employ computer-simulated situations to train and/or to test employees. A well-known example of this is the flight simulator used by airlines to test and train pilots.

Natural languages

Another aspect of AI is concerned with providing computers with the ability to communicate in a human language such as English, French, or German. Much research has been going on in this field with some recent successes being reported.

Vision systems

It is possible to use computer-controlled devices to visually inspect items on a factory production line to determine their suitability for their intended purpose. A digital photograph is taken of each item and this is turned into data that can be compared with other data, held in memory, that represents a perfect item. If the two sets of data differ from one another it means that the inspected item is not perfect and the computer will have been programmed to then take some specific action.

Robotics

Robots are now commonly employed in factories to perform a wide variety of tasks that are repeated many times but must be carried out with great precision. Essentially there are three kinds of robot. Firstly, there are those for which every part of the task to be performed has to be carefully calculated beforehand and all the data fed into the robot. Secondly, a more sophisticated robot can be taken through the sequence of activities it will be required to follow while the computer records and stores the information about robot positions and movements. The computer is then able to control the robot to perform that task. Thirdly, if the computer system is provided with AI the robot is able to 'see' and hence learn for itself many, if not all, of the steps involved in a particular task.

5 IT in the office

Many applications of information technology are employed in the modern office. Letters, invoices, and documents of all kinds are prepared with the aid of a word processor. Information may be stored, and later retrieved, in a database and calculations of various kinds performed with the aid of a spreadsheet. Among the many facilities with which a modern office may be provided are telephones and telephone switchboards, electronic typewriters, duplicators, copiers, adding machines, dictation machines, computers, databases, videotex services, fax, and electronic mail. All of these services rely heavily upon modern electronic circuitry.

It is usually taken for granted that the term 'information technology' when applied to office practice means the use of a computer and/or computer-controlled equipment. The introduction of the personal computer (PC) has led to the widespread introduction of information technology into offices. The PC's relatively low cost means that computers may be purchased without much, if any, need to gain authorization from some higher authority, and it is able to deliver whatever the customer wants since, with suitable software, it can provide solutions to specific problems such as preparing accounts, and paying salaries and wages.

In the larger offices the PCs that are employed will probably be networked allowing all users to have access to common facilities such as printers and databases. A really large office may provide terminals that are connected to a mainframe computer, or to a minicomputer, to give a user access to all the facilities offered by a large computer.

Relative merits of typewriters and word processors

The basic mechanical, or manual, typewriter, which produces text by having small hammers strike an inked ribbon against paper, would nowadays be employed only for occasional use in the home. Electronic typewriters are able to produce excellent results with the minimum of effort. All typewriters are able to print a certain number of characters per inch at a certain speed. The **pitch** of a typewriter is the number of characters it produces per inch. For most models the pitch is either 10 or 12 but a few models are able to operate with a pitch of 15 while with some machines the pitch can be varied. The print speed varies, according to the model, between 10, 12, 15, or 18 characters per second. Usually, the faster the print speed the worse becomes the print quality. In most models the typestyles and typefaces can be altered to change the presentation of the work. The variation might well be from italic and condensed typestyle to Roman and Sanserif typeface, or **font**. Most models are able to provide centring, underlining, bold print, and superscript/subscript.

An electronic typewriter may have the following features:

1. **Correction memory**.The internal memory of the typewriter is able to store up to three lines of text, depending upon the particular model. If an error is made a pointer can be moved to the error, which can then be deleted and replaced with the correct word.
2. **Text memory**. Some electronic typewriters have the facility to store a certain amount of text. Typically, the memory capacity varies from 6 kbytes (approximately three A4 pages of text), to about 24 kbytes (approximately twelve A4 pages of text).
3. **Liquid crystal display**. A liquid crystal display (LCD) is provided with text memory to allow the user to view, and if necessary, edit text easily.
4. **Justification**. Justification means giving the text a straight right-hand edge and a straight left-hand edge by expanding some of the spaces between words. Justification is often used since it makes the text look very much neater; justification would be a very laborious task on an ordinary mechanical typewriter.
5. **Spell checker**. Some machines are provided with an internal spell checker, and this may be used to ensure that a document is free from spelling errors. It saves time that might otherwise be spent checking spellings in a dictionary.
6. **Decimal tabulation**. The decimal tabulation facility allows the user to align numbers automatically at the decimal point.
7. **Word-processor features**. Some of the more expensive models of electronic typewriter are provided with some word-processing features such as copy, move, paste, and delete, blocks of text.

Usually an electronic typewriter has a daisy wheel print head which gives good **letter quality** (LQ) print.

Printers

A printer is used to produce a hard copy of the output of a computer. Some printers produce text one character at a time while others produce complete lines of text at once. Presently the most commonly used type of printer is probably the dot-matrix printer but the ink-jet printer is increasingly employed. Offices that have considerable amounts of data to print may go to the expense of a laser printer, which gives high-quality print. The printers used in conjunction with a mainframe computer are required to print out large quantities of data very rapidly and are usually either large laser printers or line printers. Most printers are bidirectional in their operation; this means that the printing mechanism is able to move from left to right or from right to left when printing.

Daisy-wheel printer

A **daisy-wheel printer** prints characters on paper by striking a carbon, or inked, ribbon onto the paper. The characters are mounted on the tips of a number of flexible spokes which extend from the hub of a circular **daisy wheel**. The basic construction is shown in Fig. 5.1. When a particular character is to be printed the daisy wheel rotates to place that character in the uppermost position when it is next to the ribbon. Once the character is in that position a small hammer strikes it to push it against the ribbon. This impact leaves an imprint of the character on the paper. The print head is mounted on a carriage which moves the head horizontally along the line, in either direction, to the position of the next character to be printed. The striking action makes it possible to use carbon paper to obtain copies of the printed text. If a different set of characters is required it is necessary to change the daisy wheel.

A daisy-wheel printer is noisy and is relatively slow to operate, typically about 50 characters per second, although some models are somewhat faster, but it produces high-quality print. This kind of printer is more or less obsolete as far as new models are concerned but there are still very many of them in use.

Dot-matrix printer

A dot-matrix printer is used when the print quality is of lesser importance than cost. Figure 5.2 shows the basic construction of a dot-matrix printer. The print head, containing a column of pins (sometimes called needles), is moved across the width of a sheet of paper. An inked ribbon is moved slowly in the gap between the print head and the paper. Any one, or more, of the pins can be caused to strike against the ribbon to press it against the paper. Each impact of a pin produces a dot on the paper and any desired image can be created as a pattern of these dots. When a column of dots has been created on the paper the print head is moved horizontally and then

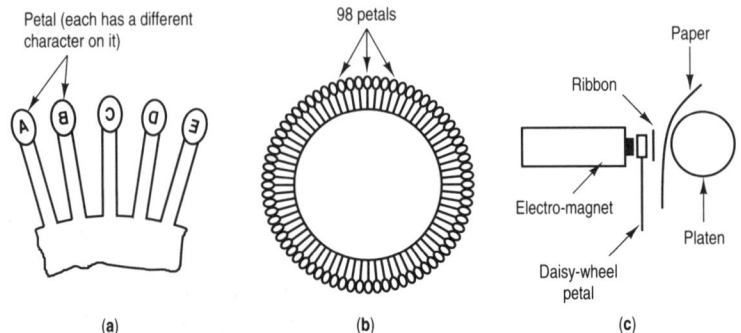

Fig. 5.1 Daisy-wheel printer.

(a)   (b)   (c)

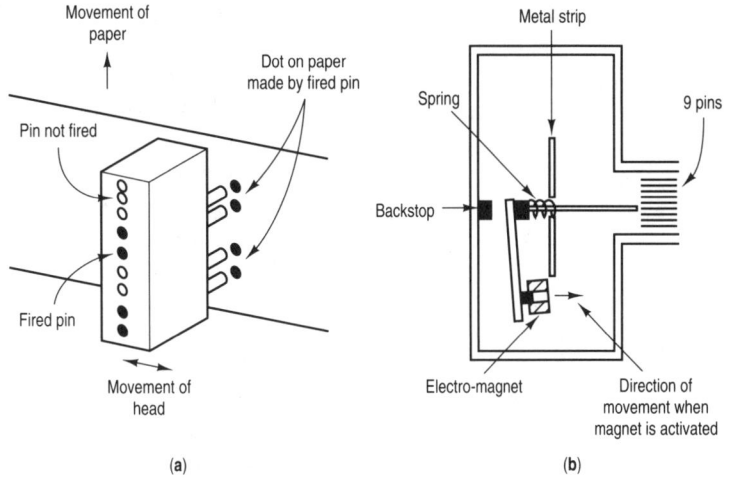

Fig. 5.2 Dot-matrix printer.

(a)

(b)

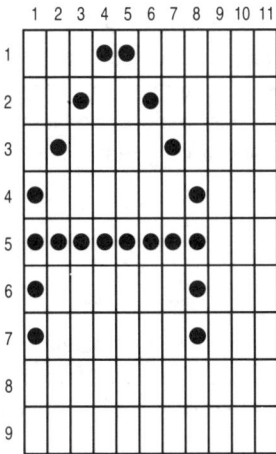

Fig. 5.3 9 × 7 dot matrix.

another combination of the pins is made to strike against the ribbon to produce another dot pattern. By moving the head and striking selected pins so that they protrude from the rest any character can be formed. In addition the characters may be enlarged, or reduced, or printed as bold, or in italics. Since characters are formed by allowing different pins to strike the paper a dot-matrix printer can also handle a variety of fonts and typestyles and is also able to print graphics.

A 9 × 7 matrix of dots is shown in Fig. 5.3; this shows how the letter A is made up from a number of dots. The gaps between adjacent dots mean that the print quality is not very good but the greater the number of dots, the better the quality. The number of vertical pins employed is generally 9 or 24, depending upon the model. The matrix for a single character may be 9 × 7, 9 × 9, etc. up to 24 × 24 for the more expensive printers. A 24-pin dot matrix printer tends to be more expensive than a 9-pin machine.

Near letter quality (NLQ) print may be produced with a 9-pin machine by double printing, although a 24-pin printer will always give NLQ results. With double printing, one line of the text is printed and the print head is then moved either downwards or sideways, a very short distance, and the line is printed again. This action fills the gaps between the dots forming each character, thus improving the print definition, but the action is at the expense of a reduction in the printing speed.

Typically, the printing speed of a dot-matrix printer might be up to 400 characters per second (or more on expensive models) for draft quality print, and up to about 120 characters per second for NLQ print. The dot-matrix printer is very noisy in operation but it is cheap to both buy and run. Since an impact printing mechanism is employed, carbon copies of the printed output can be obtained.

Ink-jet printer

An **ink-jet** printer is not an impact type. Instead, characters are formed by spraying the paper with tiny droplets of a fast-drying ink as the print head is moved horizontally. The print head contains a number, typically 64, of very fine ink tubes, or nozzles, through which the ink droplets are fired. The ink jet is activated by a pump that builds up pressure that forces the ink out of a tube and onto the paper. The idea is shown in Fig. 5.4. The sprayed ink produces a stream of inked dots on the paper. The characters are not formed immediately the print head passes but are gradually built up as the head moves back and forth along the line. Since each character is formed from hundreds of these tiny dots the print quality is very good and graphics may be printed. The printer may be programmed to change typestyles while printing is still in progress, and is very quiet in its operation. Special paper that is quickly able to absorb the ink sprayed on to it must be used; ordinary printer paper is not suitable.

A **bubble-jet** printer operates in a similar manner but instead of using a pump the ink is sprayed onto the paper by the steam pressure created by heating the ink tube.

Ink-jet and bubble-jet printers are more expensive than either the daisy-wheel or the dot-matrix printer but their reliability is better since they have very few moving parts. Because these are not impact type printers carbon copies cannot be obtained. The main disadvantage of an ink-jet printer is that the maximum speed of printing is only about 200 characters per second.

Line printer

The **line** printer has been generally used in conjunction with mainframe computer data-processing installations where there is a need for large amounts of data to be printed at high speed. Nowadays the line printer

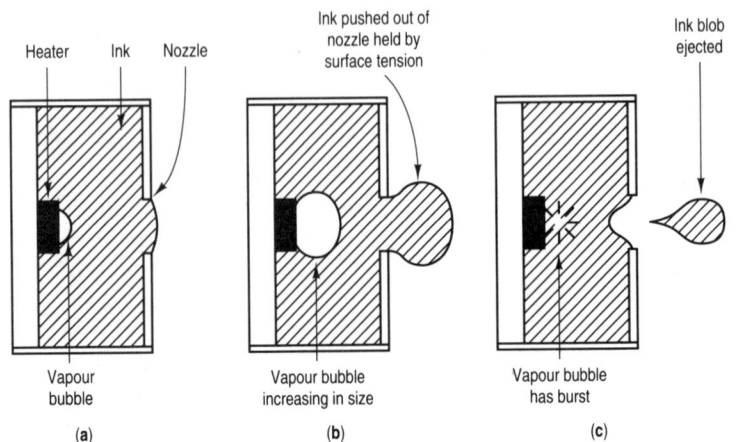

Fig. 5.4 Ink-jet printer.

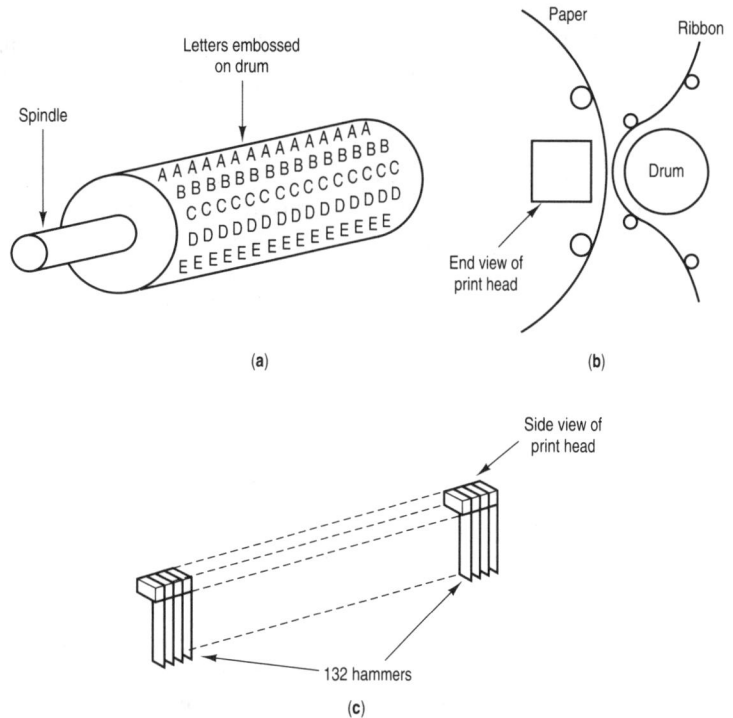

Fig. 5.5 Line printer.

is increasingly being replaced by the laser printer. The basic construction of one type of line printer is shown in Fig. 5.5. The printer consists of a revolving drum, or barrel, on the surface of which are embossed complete lines of the same character; all the characters that are available to be printed are embossed. The print head comprises a large number, usually 132, of small hammers mounted so that they are able to strike the drum. The characters are provided in as many rows as there are hammers so that each hammer is mounted directly over one row of characters. The drum is rotated at a high speed, typically 2000 revolutions per minute, so that each character comes underneath a hammer for a very short period of time. The printing paper and carbon ribbon are passed between the drum and the hammers as shown. The output from the computer is held in the printer's memory and is compared with the characters as they pass underneath a hammer. When a required character is in position, a hammer is operated to strike the carbon ribbon and in so doing pushes the carbon ribbon against the printing paper; this action causes the character underneath the hammer at that instant to be printed on the paper.

At a given instant in time the same character is underneath each of the 132 hammers. If, say, the letter A is underneath the hammers initially then **all** the As in that line of text will be printed together, and **no** other characters. As the drum rotates the next letter, B, comes underneath the hammers and then all instances of B in the line will be simultaneously printed. Next, the letter C comes underneath the hammers and all Cs in the line of text are printed, and so on. When a character that does not appear in a line of text appears underneath the hammers, *none* of the hammers is operated and nothing is printed at that instant.

There are two alternative constructions that are also used for line printers. Instead of a rotating drum a line printer may use either a chain or a metal band as the medium upon which the characters are embossed.

The printing speed of a line printer is high and may easily be as high as 2000 lines per second.

Laser printer

A laser printer looks very much like a photo-copier and it uses a similar but more advanced technique. The basic construction of a laser printer is shown in Fig. 5.6. A laser beam writes characters and graphics uniformly as bit patterns onto a rotating drum. The drum has a negative electrical charge applied to its highly resistive surface. As the drum rotates it is scanned by a laser beam. The laser beam is controlled by a microprocessor within the printer and fires a burst of light at the drum at each point at which a dot is required. Wherever the laser beam strikes the drum the negative electrical charge at that point is able to flow to earth and is then lost. As the drum rotates it comes into contact with powdered ink, or **toner**, which is also negatively charged. The toner is repelled by the negatively charged parts of the drum but is attracted to the parts of the drum that have zero electrical

Fig. 5.6 Laser printer.

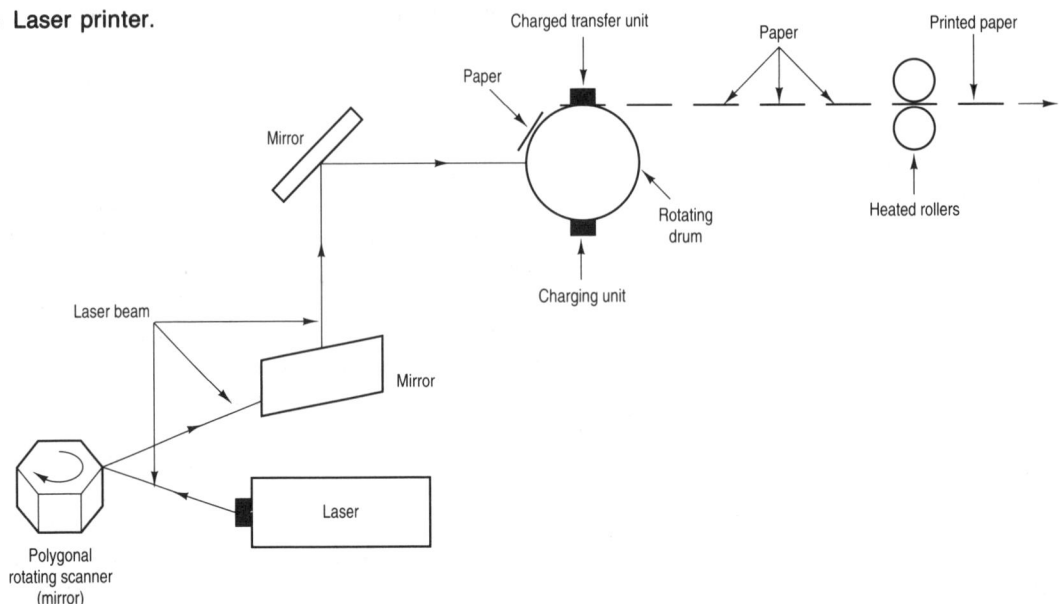

charge. As a result the toner moves on to, and adheres to, the areas of the drum that have been written to by the laser beam. By now the printing paper has been given a positive charge and, as the paper passes the drum, particles of the negatively charged toner on the drum are transferred to the surface of the paper. The required image is hence printed on the paper. Since the electrical charges have a very short lifetime the paper is passed through heated rollers which melt the toner particles and bond them firmly to the paper. Thus, characters and shapes are transferred first to the drum as electrostatic images, and then onto the paper to give a high-quality printed output.

As with a dot-matrix or ink-jet printer, the individual characters are formed on a matrix or grid. The resolution of modern laser printers is defined by the number of **pixels** that form the grid. The standard value is 300 × 300 pixels per square inch, or about 14 000 pixels per square centimetre.

Before a page of text can be printed it must be stored in the printer's memory and the printer's internal microprocessor programmed to control the printing process. Instructions to the laser printer, to let it know just where the beam should strike the rotating drum to place the dots that make up each character, are sent in a special printer language. Several such languages exist but two of the more commonly employed are known as **Printer Command Language** (PCL) and **Postscript**.

A laser printer produces high-quality print at a speed of about 8 pages per minute or more and it is very quiet in its operation. As various typestyles, typefaces, and graphics are available, these machines are used for desktop publishing. Laser printers tend to be rather expensive compared to most other types but their prices are rapidly falling.

Choice of printer

The choice of a particular type of printer for a specific application is mainly based upon such considerations as cost, print quality, printing speed, noise, and reliability. The relative merits of the different types of printer are tabulated in Table 5.1.

Table 5.1

Printer type	Speed	Running costs	Purchase price	Noise level	Main use	Carbon copies	Print quality
Daisy-wheel	low	low	low	medium	home	yes	very good
Dot-matrix	medium	low	low	high	PC	yes	poor/good
Ink-jet	high	medium	medium	low	PC	no	good
Laser	very high	high	high	low	mainframe and PC	no no	very good
Line	high	low	high	high	mainframe	N/A	good

The daisy-wheel printer is mainly used in conjunction with home computers and in small offices. It is provided, for example, with the Amstrad PCW 9512 computer/word processor. The daisy-wheel printer cannot print graphics. The dot-matrix printer is one of the cheapest to buy and it allows the user to produce reasonably good-looking text quickly and cheaply. The 24-pin versions produce better quality print than the 9-pin versions but may be more expensive. The dot-matrix printer is versatile and able to print graphics as well as text. For new installations that require low-cost printing the ink-jet printer is gradually replacing the dot-matrix since their relative purchase prices are falling each year, and this type of printer is both quiet and able to produce good-quality print. Perhaps its main disadvantage is the risk of the print smudging because the ink is not quite dry, rather like newspaper print. Not all kinds of paper can be used with an ink-jet printer if good results are sought; the use of the wrong type of paper will probably result in the ink spreading on the paper. The laser printer gives good print quality at high speed and very quietly. It combines the print quality of the daisy-wheel printer with the versatility of the dot-matrix printer and it is both faster and quieter than either. It also gives better quality print than an ink-jet printer but is, however, more expensive. This type of printer is used when the quality of the print is of more importance than cost, and laser printers are, for example, used in desktop publishing. There are really two kind of laser printer: large, very high-speed machines designed for use with mainframe computers and cheaper, less efficient versions designed for use with PCs. The line printer is only used when large amounts of data are to be produced as quickly as possible and this means that it is used in conjunction with mainframe computers, but now large laser printers are becoming increasingly competitive. The line printer does not provide a NLQ feature.

A few other types of printer are in existence, but they are much less frequently used than those described. These other printers include the crystal shutter printer, the LED printer, and the thermal printer.

Word processors

Word processing is the typing and editing of text using a computer. When a letter is typed using a manual typewriter it is easy to make some mistakes, and, although these mistakes can be corrected using Tippex or by some other means, the final result may be somewhat lacking in appearance. If the errors are many, or alterations are wanted, it will be necessary to re-type the letter. An electronic typewriter has some provision for error correction but this is limited in comparison with a word processor. A word processor can be employed for the creation and consequent editing of all kinds of text, ranging from short letters to complete books. The use of a word processor allows the user to type quickly without bothering too much about errors at the time, and then rapidly to correct any errors afterwards. The document is not printed on paper until the user is satisfied with its appearance and it may be stored on a disk. Further copies of the document can be printed out at any time. If, later, some revision is needed the document may be loaded from the disk and the required alterations made. There is no need to re-type the whole document, which saves time and hence improves office productivity.

A word-processing system may be either **stand-alone** or **communicating**. The components of a stand-alone system are a processor, a printer and a disk store, while a communicating system consists of two, or more, processors and disk stores, a switching system, and one, or more, printers. In either case the system is unable to do anything without software to drive it and hence a suitable word-processing package is necessary.

The vast majority of word processors use an IBM compatible PC. The faster the speed of a processor the quicker the word-processing software will be able to carry out such actions as going to a specified page, or performing a spelling check, in a long document. Since 80486 PCs are more expensive and very fast they are not often used as word processors, and when they are it is mainly as the controlling machine in a network, i.e. in a communicating word-processor system, although the price of 80486 machines is falling so rapidly that this situation may soon change. The 80286 is more or less obsolete although large numbers of these machines are still in use. An 80386, or higher, is required to run WINDOWS-based word-processing packages with reasonable speed. A disk drive, hard and/or floppy, is required to store the word-processing package itself and to store the text produced. The DOS software will occupy anything up to about 8 Mbyte of storage space. A hard disk with a capacity of 200 Mbytes or more is generally only used when shared access to the drive is given to a network of PCs. Expensive printers, such as a laser, will also probably have shared network access.

A large number of word-processing packages is available, but they all have certain facilities in common even though the commands necessary to use them will probably differ. Status information is generally shown at the top, and perhaps also at the bottom, of the screen. This information might include the page number, the line number, the set page width in terms of a ruler given at the top of the screen, and the pitch and the line spacing in use. The main option for a PC user is whether to use a DOS- or a WINDOWS-based package. For a DOS system the letters appearing on-screen are of fixed appearance and so the display is more or less the same as typed text. A WINDOWS-based word-processing package uses a graphics mode rather than a character mode and so the screen may show different text sizes, styles, and typefaces, which means that it is able to give (almost) WYSIWYG. This means that a WINDOWS word processor is easier to learn to use and also makes it possible to achieve a better hard copy, since the user can vary the on-screen display until the document looks exactly right. DOS word-processing packages are better for rapid text entry and are often easier to read and hence less tiring to work with.

Before text is entered at the keyboard it will be necessary to load the word-processing software and to call one or more of the command menus. A menu displays a list of available activities from which the user is able to select one. The method of selection may be by keying a particular number, or letter, or moving a highlighted choice by an arrow key and then pressing the ENTER key or, if a Windows package is used, using a mouse. Usually, to start work there is a choice of such activities as create new text, edit existing text, print existing text, or merge existing text. If a new document is to be prepared the first choice will be selected, which will probably require the user to give the file a name, and this will present the user with a blank screen except for the status information.

Non-printing characters

All word processors employ a variety of non-printing characters which control such things as carriage return and tabs which are displayed on-screen but do not appear in the hard copy, and control characters such as bold on/off and italics on/off. The latter may be permanently displayed on-screen with some packages, generally either highlighted or shown in a different colour, or may be turned on and off as required by the user. When displaying text on-screen the word processor breaks lines at the right-hand edge. The exact position of a break is set by a return character. This may be either a **hard return** or a **soft return**. A hard return must be obeyed and it is obtained whenever the RETURN or the ENTER key is pressed. In word processing a hard return denotes the end of a paragraph. A soft return is temporary and it is inserted by the program to denote the end of a line within a paragraph. If text is inserted or deleted the position of the soft return will automatically be altered to fix the new end-of-line position. Some word processors provide a means by which graphics files may be imported into a document from an outside source.

Producing a document

Creating the document

A format for the document should first be set up. Most packages offer a variety of standard formats for letters, memos, reports, etc., and, if suitable, one of these should be used. It may be required to alter the layout in some way. All, or any, of the following choices will probably be provided and can be selected from a menu: the paper type — e.g. A4, A5, 11 in. fanfold, or 2 in. labels, single sheet or continuous stationery; page layout — altering the default gaps at the top and bottom of the page to some other values, and left and/or right margin widths; tabs — left, right, or decimal;

justification — i.e. giving the text both a straight left-hand edge and a straight right-hand edge by expanding some of the spaces between words. Justified text looks very neat and is something that is not easily done on a manual typewriter (this paragraph is unjustified and has a left-hand tab; its appearance should be compared with that of the previous text); print — the pitch, the line spacing, the carriage return spacing, and the line pitch may all be varied from their default values if wanted.

The text can now be entered at the keyboard, correcting any errors noticed *en-route*. There is no need to keep an eye out for the end of each line since the software will provide **automatic word wrap**. This term means that when the cursor reaches the right-hand margin it will automatically move to the left-hand margin of the next line. Any word which will not fit into the line will be moved to the next line. Words are never split over two lines, unless a soft hyphen is inserted, and the RETURN key should not be pressed. The RETURN key should be pressed only at the end of a paragraph.

A number of commands are provided in all word-processing packages to allow the text to be enhanced in various ways; typical commands include bold print, italics, underlining, superscript and subscript. Some packages show these effects on-screen as they will appear on paper; some highlight or colour the enhanced characters; while still others show the codes used to send instructions to the printer. The ways in which text enhancement is obtained varies between different word-processing packages; to underline a word, for example, the required keying might be 'ALT u' or '+ u'.

At regular intervals, say every hour, the text should be saved just in case a power, or computer, failure should occur; some word-processor packages do this automatically.

Editing the document

Once the text has been entered it should be edited. The editing process is aided by a number of keys that allow the cursor to be moved around the screen. The labelling of these keys may vary from computer to computer but the actions provided generally include: (a) PAGE — moves the cursor to the end of the page; (b) ALT PAGE — moves the cursor to the beginning of the page; (c) END of line — moves the cursor to the end of the line; (d) ALT END of line — moves the cursor to the beginning of the line; (e) WORD — moves the cursor to the end of the word; (f) ALT WORD — moves the cursor to the beginning of the word. Many computer keyboards have keys marked as PgUp and PgDn that move the cursor vertically a page at a time. In some word-processing packages a mouse can be used to move the cursor around the screen. Also some packages allow two, or more, windows to be set up on the screen at the same time. This facility allows two parts of a document, or two different documents, to be viewed simultaneously.

The facilities generally provided to assist in the editing of a document include the following:

1. **Delete**. If the cursor is positioned on top of, or perhaps alongside, an incorrect character pressing the appropriate key (often labelled as DELETE) will delete that character. There is often an UNDO facility provided which allows a character deleted in error to be restored. Some packages allow a mouse to be used to mark the block of text to be deleted.
2. **Overwrite**. This is used if it is required to type new characters on top of characters already there.
3. **Copy/paste**. A section of the text may be copied, removed from its present position in the document and then inserted elsewhere. The procedure varies with the word-processing package but, in one package for example, the procedure is press COPY, mark out the section of text to be removed, press CUT, move the cursor to the new starting position of the shifted text, and then press PASTE. Other packages may require a block of text to be marked by 'beginning' and 'end' markers, perhaps by using a mouse, and then shifted to its new position by a MOVE command.
4. **Search and replace (find and exchange)**. These commands provide an automatic search for a word and, if also requested, for the replacement of that word with another word. The word-processing package can be asked to stop at each occurrence of the word before a key is pressed to effect the exchange or to replace each located word automatically.
5. **Word count**. Most word-processing packages provide a word count facility although in some cases this is integral with the spelling checker.

Saving the document

When the text has been edited it must be saved on to the text disk. A key is pressed which generally brings on-screen a menu of choices such as SAVE, SAVE and CONTINUE, and SAVE and PRINT. The first of these choices saves the text and then returns the user to the command page. SAVE and CONTINUE should be used at periodic intervals to ensure that the work is not lost. Should a power failure occur, the text will be stored on disk and the user returned to the point in the text reached when the command was given. SAVE and PRINT first saves the text and then produces a hard copy. The hard copy can then be checked for any errors that may have been missed and for any layout changes that may seem desirable. If any such are detected the text stored on disk can be loaded into the computer and amended until the required result is obtained, and this latest version can then be printed.

The steps that should be taken to prepare a letter, for example, are:

- Set the margins. This is usually done via a menu command under the heading of setting the layout.

- Select justified or non-justified text. This is also achieved via a menu command.
- Start to type in the name and address of the addressee and the text of the letter. Any typing mistakes can be corrected as they are noticed. The RETURN key should only be pressed at the end of each line in the address and for the end of each paragraph.
- When the letter is finished it should be saved, then printed out. One command might well perform both activities.

Most word-processing packages are WYSIWYG to the extent of showing where lines and paragraphs end, the spaces between lines and paragraphs, headings, etc., but most do not show how the various enhancements look. True WYSIWYG is only obtained with a DTP package.

Mail merge

Mail merge is the ability of a word-processing package to print out multiple copies of a standard, or **form**, letter each showing a different name and address. The user provides the basic letter, which is a common text that forms the major part of each letter to be mailed, and a mail list file that contains names and addresses and perhaps data such as an account number or current account details. Some word-processing packages import the data from a database while others require the user to create a document that contains the names and addresses. Standard letters are used in many offices, perhaps requesting payment of an overdue bill, or the rejection of a job application and so on. The standard letters can be held in one file and a list of names and addresses held in another file.

Databases

The main factors that influence the choice of the means of storage and retrieval of information in an office are: costs, both initial and running; ease of use; the storage capacity; the speed with which required information can be accessed.

The methods used to store information include: paper, folders, and filing cabinets; computerized systems, such as databases; microform, such as microfilm and microfiche. Whatever the method used the data records must be available when they are required and their retrieval must be rapid. This means that records must be stored in such a way that they are easily retrieved as and when required so that reference can be made to the stored data, and, if necessary, the data can be up-dated and/or amended. Hence the records should be organized into files and the location and contents of each file, and the order in which the records are arranged, should be known.

The use of paper records, letters, etc., that are kept in folders and then filed in filing cabinets has been in use for a very long time and is still in use today, particularly in smaller offices. This is a fairly straightforward process for a manual or paper-filing system since it merely demands that the records are visually checked to find a required file. The disadvantages of a paper document storage system are manifold; for example, manual processing involves the sorting by hand and up-dating of handwritten and typewritten letters, lists, invoices, etc., plus the use of ready reckoners and calculators to carry out calculations, and both may take considerable time and be prone to errors. If an office worker is dealing with just a few transactions it is easy to check that details are entered correctly and that the same details appear on both the record sent out and the one retained for reference purposes. If, however, a large quantity of data is to be dealt with such checking would become a laborious task, and one that would be very error prone because the office workers would become bored and tired. There would also be the problem of addressing all the envelopes and filing all the reference copies in a filing cabinet. If, at some later date, the need arises to refer to a filed record the filing system used must be efficient if the data is to be located in a fairly short time. In a large office considerable storage space may be needed to accommodate all the filed records, especially if there is a need for records to be kept for some years.

A **database** is a collection of data organized in such a way that individual items of data can easily be located when needed. A database is not necessarily a software package; for example, a telephone directory is a database in which information is arranged in the alphabetical order of the customers' surnames. A **database management system** (DBMS) consists of a number of programs that organize the way in which the database is used.

Before data can be stored in a database, the database must be informed of the nature of the data to be stored, e.g. names and addresses. All databases have some way of allowing the user to define how the data will be stored and how it will be shown on-screen. The database must also be told the kind of data that each heading represents, e.g. alphanumeric or numeric. Each database has its own methods by which the required information is inputted and some databases are much easier to set up than others. Most packages use a screen-painting tool which allows records to be designed using either the cursor keys or a mouse. This is easiest using a Windows package. A database may be either **flat-file** or **relational**. A flat file is like a card index; it is a single data file in which each record has an identical format, for example, a list of names and addresses. For simple tasks in which each piece of data is self-contained a flat-file database may be adequate. A flat-file database does not have the capability to pull-in data from other database files, so if the application has data items that appear in multiple records then the items must be entered and stored separately for each record. A related database overcomes this problem; it allows files of related information to be linked together using indexes or keys. When a record is viewed the database manager retrieves the appropriate records in the related files and pulls in the required fields. The result is that data need only be stored once, irrespective of the number of records it is included in and for which it may be required.

Databases are available using both the DOS and the Windows environments; the Windows versions require the use of a mouse. If a database is to run as a multi-user application then a network version is necessary.

Concepts

There are a number of concepts associated with databases with which a user ought to be familiar.

Files

A file is an organized collection of related items of data. This is, of course, similar to a file of papers kept in an office filing cabinet. Each file stores data of a particular kind — for example, names and addresses, or electrical component part numbers, or train fares. A database is able to store data on a variety of different topics in different files. Most database packages store data using a particular format; some may store all the data plus the index files in a single-disk file, while others may store each data and index file in a separate disk file. The most commonly employed format is ASCII, but the dBase file format is also often used.

Records

A file consists of a number of **records**. Each record contains the data relating to a single item. In a name and address file, for example, each record contains the name, address, and perhaps the telephone number of a particular person. Each record has an identical structure and holds the same kind of information.

Fields

Each record is made up of a number of **fields**. A field is the space occupied by a single item of data. In a name and address file it might be a person's first name, or surname, or address, or telephone number.

To ensure that different records are easily distinguished from one another it is usual to make sure that the contents of at least one field is different from the contents of the same field in every other record. This field is then known as the **key field**. In a name and address file the key field could hold a National Insurance number, or a tax reference number, or some other unique piece of data. If each field contains a unique reference number which appears on all the records relating to that person, then that reference is known as a **record key**.

WINDOWS databases are the easiest to use. A file can be created on-screen by drawing a box with the aid of a mouse and then moving the box to the wanted position. The mouse can then be clicked to call up a dialogue box in which the field's name, its type and size, and the validation procedure can all be defined.

Creating a new file

A data file consists of both the actual data and the structure in which the data will be entered and displayed. There is no default structure for a database and so a structure must always be set up for each new file. Some programming commands may also be included to make it possible for such quantities as totals, or averages, to be calculated automatically.

The way in which the structure of a database file is set up varies with each database but on-screen menus are provided to make the task as simple as possible. If, again, a name and address file is being created the structure chosen will probably be of the form shown in Fig. 5.7.

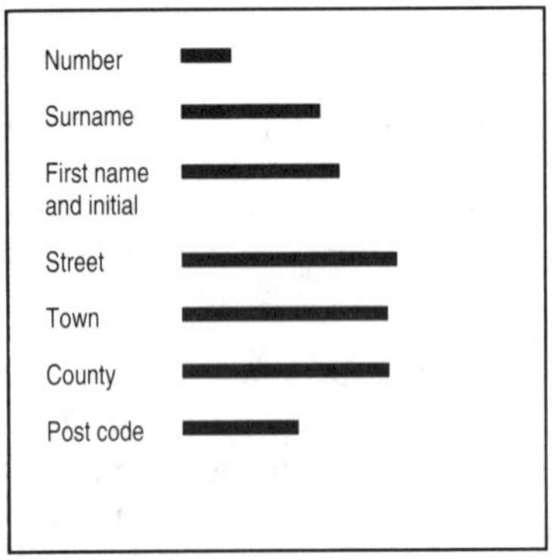

Fig. 5.7 Database structure.

Entering data

Once again the method of entering data into a database varies with the database concerned but often an EDIT command is selected from the main menu. This command then brings on-screen a display similar to that shown in Fig. 5.7 along with another menu. One of the choices given by this new menu should allow new data to be entered into the database. Once this has been done a file entry will look somewhat like that shown in Fig. 5.8.

1	1001
2	WILLIAMS
3	PETER R
4	22 LOW STREET
5	OLDHAMPTON
6	WESSEX
7	XA5 3WQ

Fig. 5.8 Name and address file in a database.

Some database packages will allow new fields to be added to an existing structure while others will not. Data validation is used to ensure that the entered data is accurate. The most basic version of a validation procedure is that the fields are defined as being integer, or data, or money, etc. types, and an error is signalled if an inappropriate value is entered.

Calculations

A database has the facility to allow simple calculations to be performed upon each numeric record and for the results of these calculations to be displayed in each record alongside the other data. The calculations are carried out using **formula fields**, which contain simple programming instructions that enable the computer to carry out mathematical and logical operations on the contents of records held in the database. The formula for a wanted operation must be entered by the user. A simple example is to multiply the contents of a field by 1.5: the formula would be [field number] $*$ 1.5. Similarly, to divide the contents of a field by 1.5 the formula used would be [field number]/1.5.

Some databases include a facility that allows a calculation to be applied to more than one record at a time.

Interrogating a database

One of the characteristics of a database is that it is possible not only to retrieve information in the same form as that in which it was originally inputted, but also to manipulate the information in a variety of ways. A **query** facility allows the user to specify certain fields and to then view the results on-screen. Examples are many and include (a) searching for records that match certain specifications, (b) sorting records into a different order from that in which they were originally entered into the database, (c) selecting the format and the specific fields that are to be printed out, thus creating a **report** of a file.

Searching for records

When data is to be retrieved from a database the search criteria are typed in at the keyboard. The computer will then search for, and display, the wanted record(s). A database program always includes an automatic search facility. To speed up the retrieval process it is usual to **index** the fields. Indexing produces and stores lists of the records sorted on the indexed fields. The computer is able to use these lists to locate, rapidly, any wanted record. An index is automatically up-dated whenever a record is either added or deleted from the database. Indexing does have the disadvantage of occupying some disk storage space and if the index is large the speed with which records are amended will be reduced because of the time taken by the computer to up-date the indexes.

Sorting records

It is often necessary to be able to sort the contents of a database into a different order from that in which the records were originally entered. The on-screen menu will include a SORT choice and this should be selected. Another menu will then probably appear which asks the user to state the kind of SORT required, e.g. sort surnames, or addresses, or towns. If a name and address file is again considered, should the SORT be in ascending order or descending order? Pressing the correct key will then cause the SORT to be carried out.

Print-out

It is usually possible to obtain a hard copy of all the details of a single record, or of all the records in a file one after the other. Most databases come complete with a **report generator** which allows the user to produce reports such as lists, mailing labels, and form letters.

Typical information stored in a database

Stock records

A business can keep track of all purchases, issues of spare parts, materials held in stock, stationery, etc. The database can be programmed to order more stock automatically when the stocks of various items fall below some predetermined level. Taking a new item into stock would involve recording the stock's description, size, and weight and this information would remain constant all the time the item is held in stock. Also recorded, but subject to change, would be the quantity of stock held and the unit price.

Purchase records

A business may well need to keep records of the suppliers of various items and of the materials purchased, their prices, and order details. When a requisition is received from the stock record system it is used to generate a purchase order which can then be sent to the appropriate supplier. The system can also keep a track of outstanding orders and check that the received supplier's invoices are correct.

Customer records

A business will keep records of all its customers' names and addresses, details of orders they have made and of payments received. The database can be used to keep a note of outstanding bills and to generate reminders.

Personnel records

A business will keep records of its employees, their names, addresses, birthdays, qualifications, experience, positions held in the company, perhaps reports on their work by superiors, and so on.

A multi-user database needs a feature, known as **file/record locking**, which prevents two, or more, users trying to up-date the same file at the same time.

Spreadsheets

A spreadsheet organizes data in the form of a table known as a worksheet. Essentially a spreadsheet consists of a large number of boxes, or **cells**, arranged in vertical columns and horizontal rows. Each column corresponds to a field in a database and each row corresponds to a database record. Spreadsheets are widely employed in both commerce and industry to assist in making decisions, when a number of 'what if' options are tried out and the results are immediately displayed. A typical use for a spreadsheet is to keep track of the incomings and outgoings of money to and from a bank account, and an example of this is given later.

Each cell has a unique address and may have placed into it textural data, numeric data, or a formula that specifies a rule by which a cell's value is to be calculated. A spreadsheet always contains many more cells than can be displayed on-screen at one time and so the screen display ought to be treated as a window which can be moved over any desired part of the sheet. The size of a spreadsheet is limited by the capacity of the computer's internal RAM; this is in contrast with a database whose size is limited by the available external disk capacity.

Normally the cells in a spreadsheet are uniquely identified by a combination of letters and numbers. The columns are labelled with letters, the left-most column being labelled as A, the next column to the right as B, and so on right through to Z. After the 26th column the labelling continues with AA, AB, AC, and so on. Rows are labelled by numbers starting from 1, followed by row 2, and so on to the highest numbered row in the sheet. The cell in the top left-hand corner of the spreadsheet is labelled as cell A1, the cell just to the right of it is cell B1, the cell immediately below cell A1 is cell A2, and so on.

A spreadsheet consists of a grid of such cells and it allows the user to write data into any of those cells. The data may be an item of text, a numeric value, or a formula which uses values held in cells elsewhere in the spreadsheet. An entered formula is not visible in its cell, although it may be visible at the top of the spreadsheet, but it enables calculations to be made on the contents of other cells. The result of such a calculation appears in the cell containing the formula. In some spreadsheets the computer is told that a formula is to be entered next by pressing a specified key, in others the formula is preceded by a mathematical operator such as + or ×.

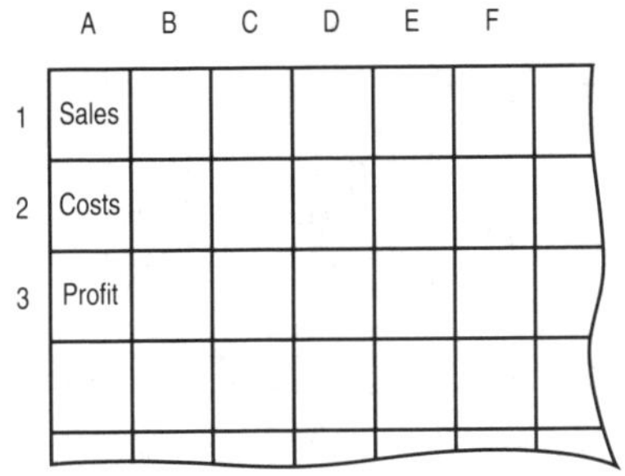

Suppose that the profit to be made from selling certain electronic devices is to be calculated. The profit will be equal to the revenue obtained from selling the devices minus the total costs of the selling operation. The basic spreadsheet structure would be as shown in Fig. 5.9. Three labels — sales, costs, and profit — have been inserted into the cells A1, A2, and A3. Cell B1 will have the sales figure inserted into it, and cell B2 will have the figure for costs inserted. Into cell B3 will be inserted the formula that will calculate the profit made; clearly the profit made is equal to the difference between the sales price and the costs involved in the selling. Hence the formula to be inserted into cell B3 is [B1 − B2]. When figures are inserted into cells B1 and B2 the profit figure will automatically appear in cell B3. Thus if the selling price of the equipment is £3000 and the total selling costs are £2500 the on-screen display would be that shown in Fig.

Fig. 5.9 Basic spreadsheet structure.

Fig. 5.10 Spreadsheet calculates profit.

5.10. If, now, the selling costs were to be reduced to £2400 and this new figure is entered into cell B2, the profit figure in cell B3 will immediately change to £600.

The formulae that may be entered include the mathematical operators, plus ($+$), minus ($-$), multiply ($*$) and divide ($/$), financial functions such as depreciation and discounted cash flow, statistical functions such as mean and standard deviation, trigonometric functions, and many more.

This basic concept of a spreadsheet can be extended to allow the profit for a complete year to be determined. The extended spreadsheet is shown in Fig. 5.11. The formula shown would not be visible in the cells but may be shown at the top of the screen when the cursor is placed in a cell. If any of the monthly sales and/or costs figures were to be altered the corresponding figure for the monthly profit would also automatically alter to its new value. Figure 5.12 shows a typical display for such a spreadsheet.

The spreadsheet could now be even further extended so that it is able to calculate the total figures for sales, costs and profits month-by-month and for a complete year. The structure of the spreadsheet would then need to be further altered, as shown in Fig. 5.13. The labels for the new rows now occupy two cells and so the January label

Fig. 5.11 Spreadsheet to calculate profit over one year.

Fig. 5.12 Typical spreadsheet display.

	A	B	C	D	E	F	G	H	I	J	K	L	M	N
1			Jan.	Feb.	March	April	May	June	July	Aug.	Sept.	Oct.	Nov.	Dec.
2	Sales													
3	Costs													
4	Profit		C2–C3	D2–D3	E2–E3	F2–F3	G2–G3	H2–H3	I2–I3	J2–J3	K2–K3	L2–L3	M2–M3	N2–N3
5	Running sales			C2+D2	D5+E2	E5+F2	F5+G2	G5+H2	H5+I2	I5+J2	J5+K2	K5+L2	L5+M2	M5+N2
6	Running costs			C3+D3	D6+E3	E6+F3	F6+G3	G6+H3	H6+I3	I6+J3	J6+K3	K6+L3	L6+M3	M6+N3
7	Running profit			D5–D6	E5–E6	F5–F6	G5–G6	H5–H6	I5–I6	J5–J6	K5–K6	L5–L6	M5–M6	N5–N6

Fig. 5.13 More detailed spreadsheet.

has been moved to cell C1, the February label to cell D1 and so on. As before the formulae would not be displayed in the cells. The total sales for January and February are calculated by adding the individual monthly sales so the formula inserted into cell D5 is [C2 + D2]. The total sales for January, February, and March are obtained by adding the March sales figure to the sum of the January and February sales, hence the necessary formula is [D5 + E2]. The remainder of the year's sales are calculated using similar formulae, e.g. for April the formula is [E5 + F2].

The running costs at the end of February is the sum of the costs incurred for both January and February and hence the required formula is [C3 + D3]. The total costs at the end of March are equal to the sum of the total costs at the end of February and the March costs, so the formula inserted into cell E6 is [D6 + E3]. In the same way the formula that must be inserted into cell F6 is [E6 + F3]. Lastly, the total profit made after the first two months of the year is the difference between the total sales and the total costs, i.e. the difference between the values stored in cells D5 and D6. This means that the formula that must be inserted into cell D7 is [D5 − D6]. The remaining formulae for profits in the following months of the year are similarly calculated.

It is quite possible to work through the columns and rows of the spreadsheet entering a slightly different formula into each cell. This process would, however, be both slow and prone to error. Any errors made in inserting the formulae would result in the spreadsheet giving incorrect results and any such errors may well be difficult to locate and correct.

To overcome this difficulty most, if not all, spreadsheets allow the user to **replicate**, or copy, either a single cell or a complete column or row. A copy may be either **absolute** or **relative**. An absolute copy copies exactly the formula held in a cell into another cell. A relative

	A	B	C	D	E	F	G	H	I	J	K	L	M	N
			Jan.	Feb.	March	April	May	June	July	Aug.	Sept.	Oct.	Nov.	Dec.
1	Sales		3000.00	2800.00	3200.00	3400.00	4563.00	4480.00	5020.00	5310.00	4020.00	3400.00	2500.00	1800.00
2	Costs		2500.00	2550.00	2610.00	2580.00	2610.00	2600.00	2650.00	4500.00	2540.00	2420.00	2400.00	2200.00
3	Profit		500.00	250.00	290.00	820.00	1953.00	1880.00	2370.00	810.00	1480.00	980.00	100.00	−400.00
4	Running sales			5800.00	9000.00	1.24E4	1.70E4	2.14E4	2.65E4	3.18E4	3.58E4	3.92E4	4.17E4	4.35E4
5	Running costs			5050.00	7660.00	1.02E4	1.28E4	1.54E4	1.81E4	2.26E4	2.51E4	2.76E4	3.00E4	3.22E4
6	Running profit			750.00	1340.00	2160.00	4113.00	5993.00	8363.00	9173.00	1.07E4	1.16E4	1.17E4	1.13E4

Fig. 5.14 Showing the figures from Fig. 5.12 entered into the spreadsheet of Fig. 5.13.

copy preserves the nature of the formula but up-dates the actual cell references used. Thus, if a relative copy of the formula in cell C4, i.e. [C2 − C3] is taken and the cursor is then moved to cell D4, pressing the appropriate key will cause the formula [D2 − D3] to be copied into that cell. If, then, the cursor is moved on to cell E4 and the same key is pressed, the formula [E2 − E3] will be copied in, and so on. If, instead of the cursor being moved from cell C4 to cell D4 it is moved to cell C5, then the copied formula would be [C3 − C4]. This relative copying process can be used to insert the formulae into the remaining cells not shown filled in Fig. 5.13.

If the same figures are used as in Fig. 5.12, the appearance of the spreadsheet is as shown in Fig. 5.14.

Spreadsheets have found their main applications in three different areas:

- Calculations that involve large numbers of figures which need to be tabulated, added, subtracted, etc.
- 'What if' exercises in which varying data is entered into a spreadsheet to see what the effect will be upon some outputs of interest.
- The production of invoices.

Example 5.1

Derive a spreadsheet to maintain household accounts for the first six months in a year. The categories of expenditure are mortgage or rent, utilities (electricity, gas, and water), housekeeping (food, drink, and cleaning materials), car (insurance, petrol, and servicing), clothes, and miscellaneous (holidays and entertainment).

Solution

The columns can be labelled, starting with column D, January, February, March, April, May and June. Column J can be labelled as 'Total'. The cells A3, A4, A5, A9, A13, A17, and A18 can then be labelled with the main categories of expenditure. Next, the more detailed outgoings can be entered into column B.

At the bottom of each monthly column the total spending for that month is to be displayed; hence enter into cell D21 the formula [SUM(D4 : D20)]. Label row 21 as 'Total spending'.

The monthly income must now be entered into the spreadsheet. Row 23 is labelled as 'Income'. The net saving (hopefully!) for each month is the

	A	B	C	D	E	F	G	H	I	J	K
1				January	February	March	April	May	June	Total	
2											
3	Expenditure										
4	Mortgage										
5	Utilities										
6		Gas									
7		Electricity									
8		Water									
9	Housekeeping										
10		Food									
11		Drink									
12		Cleaning									
13	Car										
14		Insurance									
15		Petrol									
16		Servicing									
17	Clothes										
18	Miscellaneous										
19		Holidays									
20		Entertainment									
21	Total spending										
22											
23	Income										
24	Saving										

Fig. 5.15 Example 5.1: spreadsheet.

difference between the income figure and the total spending. To display this, label row 24 as 'Saving' and enter the formula [D23 — D21] in cell D24. The appearance of the spreadsheet is shown in Fig. 5.15.

The formulae used for January can now be replicated to produce corresponding formulae for each of the other months, e.g. [SUM(F4 : F20)] and [F23 — F21]. The total expenditure over the six months for each category can be obtained by adding up the relevant rows. Hence insert into cell J4 the formula [SUM(D4 : I4)] to display the total spending on the mortgage. This formula can then be replicated to display the spending on the other items. Lastly, the total expenditure, the total income, and the total saving for the six months can be obtained from the sum of either all the row sums or all the column sums.

Example 5.2

An electronic firm wishes to either buy or lease some expensive pieces of equipment. The capital cost of an equipment is £X with annual servicing costs of £Y. The cost of leasing the equipment is £Z per month inclusive of all necessary maintenance. If the equipment is expected to be used for five years determine which of the two alternatives will be the most economic for an interest rate of I%. The value of I will probably vary year by year. Assume that the value of the equipment is zero at the end of the five-year period.

	A	B	C	D	E	F	G	H	I	
1				Year 1	Year 2	Year 3	Year 4	Year 5		
2										
3	Purchase			Enter here	0.00	0.00	0.00	0.00		
4	Maintenance			Enter here	Enter here	Enter here	Enter here	Enter here		
5	Total cost			D3+D4	E3+E4	F3+F4	G3+G4	H3+H4		
6	Interest	rate		Enter here	Enter here	Enter here	· Enter here	Enter here		
7										
8								·		
9										
10	Interest	lost		D5*D6/100	[D11+E5]*6/100	[E11+F5]*6/100	[F11+G5]*6/100	[G11+H5]*6/100		
11	Total cost			D5+D10	E5+E10	F5+F10	G5+G10	H5+H10		
12										
13	Overall	total cost							SUM (D11:H11)	
14										
15										
16	Leasing	cost								
17		per month	Enter here							
18		over 5 years							C17 * 60	
19										
20										
21	Cost	difference							I13 – I18	
22										

Fig. 5.16 Example 5.2: spreadsheet.

Solution

If the equipment is purchased its total cost will be £$[X + Y]$ in the first year, £Y in each of the next four years, plus the interest lost on the capital sum spent and on the maintenance charges paid. The cost of leasing an equipment is £$12Z$ for each year or £$60Z$ in total. The spreadsheet is shown in Fig. 5.16. The formulae used for the various calculations are shown in the relevant cells although, in practice, they would not be visible but instead the result of each calculation would be displayed.

Different values of the purchase price X, the servicing charge Y, and the leasing cost Z can be entered into the spreadsheet. If, then, various values of interest rate $I\%$, not necessarily equal for each year, are entered the difference in the total cost of each alternative method of obtaining the equipment will be displayed.

Data protection

The data held in a database or in a backing storage system is valuable and some means must be used to protect the data from loss or damage. This is known as **data security**. Much data is of a confidential nature and requires to be protected from being accessed, and possibly altered or mis-used, by unauthorized users. This is known as **data privacy**.

Data security

A lot of time and money may have been invested in the creation of data files and it could well be disastrous to an organization if they were lost or damaged in some way. It is the usual practice to make back-up copies of files so that if the working copy should be damaged another copy is available. This backing-up is carried out periodically onto backing storage while the file is in use. To guard against theft, copies of the files are often kept in another office or even, perhaps, in another building. The copy, or copies, may be stored on a hard or floppy disk or on magnetic tape. The disks and/or tapes should be stored where they are not exposed to magnetic fields and the temperature and humidity are not high. Care should be taken in the handling of the disks or tapes since they can easily be damaged by a small scratch on the recording surface, for example. Care should be taken that a disk or tape holding valuable data is not inadvertently written over with new data; both tapes and disks are supplied with gadgets that eliminate this risk if they are properly used. This protection is very similar to the means adopted with video cassettes to ensure that a wanted recording is not erased by a new recording. Another method of protecting data that may be used is the use of generations of files. Whenever a file is up-dated the new file is called the **son** and the file from which it was produced is known as the **father**. When, at some later date, the son file is itself up-dated it then becomes the father file and the old father file becomes known as the **grandfather** file. If the son file should be lost, or incorrectly up-dated, the data can always be recovered from the father file. It is not usual to retain more than three generations of a file.

Data privacy

The great increase in the use of computers to process data has meant that much confidential data is held in databases and files of all kinds. Much of this data is of a personal nature, e.g. health, National Insurance, bank and/or building society account, and tax information, that ought not to be disclosed to unauthorized persons. Files are also held by the Police National Computer on people with criminal records, owners of cars, etc. It is therefore necessary to be able to restrict access to the information held in a database to those persons who are authorized to have that information. One commonly used method of achieving this involves the use of **user identifications** (ID) and **passwords**. ID is the name and/or number by which a user is identified to the computer system. A password is a set of characters that the computer has been previously told is connected with a particular ID. The password should be kept secret by the user and is not displayed on-screen when inputted by the user. To gain access to certain files a user must input both the correct ID and password.

An area of concern is that information about people may be held in a database about which they know nothing. One example of this is a person's credit rating. To give people some measure of protection against having incorrect information held on them, the **Data**

Protection Act was passed in 1984 which requires all organizations that operate a database to register with the **Data Protection Register**. People can complain to the registrar if they feel that their rights to privacy have been invaded, or can apply to have details of the information held about them to be provided to them. If they feel that the data is incorrect, they can apply to the registrar to have it altered or deleted. The act requires that the data held on an individual must have been legally obtained, used for a specific legal purpose, and only disclosed to third parties in accordance with such a purpose. The data must be accurate, kept up-to-date, and guarded against unauthorized access and loss.

6 Analogue and digital signals

A current or a voltage may be either **direct** or **alternating**. A direct current is one that does not change its direction of flow even though its magnitude may vary. Similarly, a direct voltage does not change its polarity. It is common practice to say, or write, d.c. current and d.c. voltage when direct current or voltage is meant, even though, strictly, d.c. current means direct current and d.c. voltage means direct current voltage. Some examples of d.c. voltages are given in Fig. 6.1. Figure (a) shows a d.c. voltage whose magnitude is unchanging with time; an example of this would be the voltage at the terminals of a battery until it neared the end of its life. In figure (b) the magnitude of the voltage can be seen to fluctuate with time, but since the polarity is always positive it is still a d.c. voltage. Two other examples are shown in figures (c) and (d); the voltage shown by (c) is known as a **sawtooth** waveform and it is commonly used in the timebase sections of television receivers, computer monitors, and cathode ray oscilloscopes. Figure (d) shows an example of a **digital** waveform of the kind that is used in a computer or microprocessor to represent a binary number or an ASCII character; because the waveform has only one polarity it is said to be **unipolar**. Lastly, figure (e) shows a triangular waveform.

An alternating current is one whose direction of flow is continually changing with time. It flows in one direction during its first half-cycle, and then in the opposite direction during its second half-cycle. Similarly, an alternating voltage has its polarity changed in each half cycle of its waveform. When a complete cycle has been completed another cycle is started and this process continues for as long as the alternating current or voltage exists. It is usual to refer to an alternating current as an a.c. current and an alternating voltage as an a.c. voltage. Two examples of a.c. voltages are shown in Fig. 6.2. Figure 6.2(a) shows a **sinusoidal waveform**; this is a very important waveform in electrical engineering for several reasons. It is the waveform of the mains supply voltage supplied by the electricity companies to the home, the office and the factory; most oscillators used in electronic equipment produce an output of sinusoidal waveform; many tests and measurements are carried out using sinusoidal signals; and, lastly, any repetitive signal, no matter what its waveform, consists of the

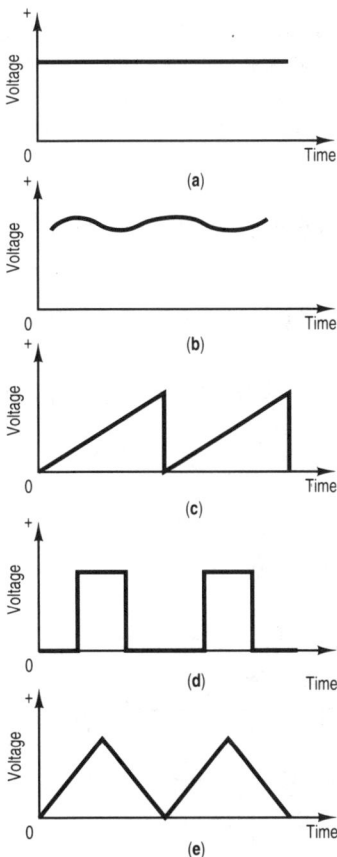

Fig. 6.1 D.C. voltages: (a) steady; (b) fluctuating; (c) sawtooth; (d) unipolar digital; and (e) triangular.

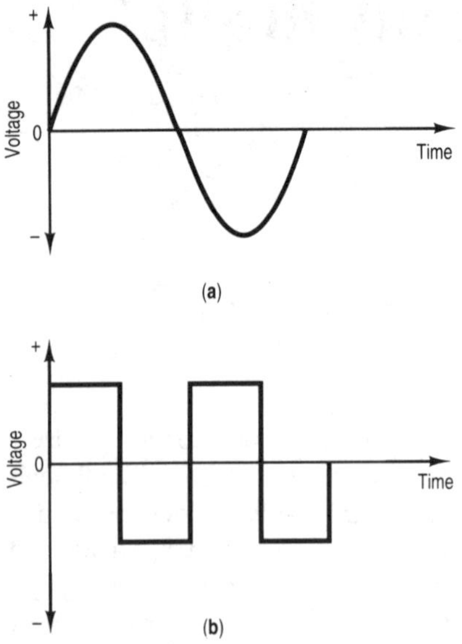

Fig. 6.2 A.C. voltage: (a) sinusoidal; (b) bipolar digital.

sum of a number of sinusoidal waves. Figure 6.2(b) shows a **bipolar** rectangular waveform; this type of signal is not used in a computer but it is used for the transmission of digital signals between two computers, or between a computer and a peripheral.

The rectangular waveforms shown in Figs 6.1(d) and 6.2(b) are, respectively, examples of d.c. and a.c. voltages. A fluctuating d.c. voltage may always be regarded as consisting of an a.c. voltage superimposed on a steady d.c. voltage whose magnitude is equal to the mean or average value of the fluctuating d.c. voltage. This concept is illustrated in Fig. 6.3. Figure 6.3(a) shows a fluctuating d.c. voltage. It is equal to the sum of the steady voltage in Fig. 6.3(b) and the bipolar voltage in Fig. 6.3(c). In Figs 6.3(a) and (c) the **periodic time** of the waveform is the time occupied by one complete cycle of voltage changes; alternatively it may be stated as being the time interval between the leading edges of two consecutive pulses, or between two consecutive trailing edges. The periodic time T is shown labelled in the figure. The frequency of each waveform is equal to the reciprocal of the periodic time.

Each cycle of a rectangular waveform may be divided into two parts, known as the **mark** (M) period and the **space** (S) period. These periods are marked on figure (c) and can be used, in two different ways, to give an indication of how far a rectangular waveform is from square.

Mark–space ratio

The **mark-to-space** ratio of a rectangular waveform is given by:

$$\text{Mark-to-space ratio} = \text{Mark/Space} = t_1/t_2 \qquad (6.1)$$

An alternative term that is also often used is known as the **duty factor**, or the **duty cycle**.

$$\text{Duty factor} = \text{Mark/(Mark} + \text{Space)} = t_1/T \qquad (6.2)$$

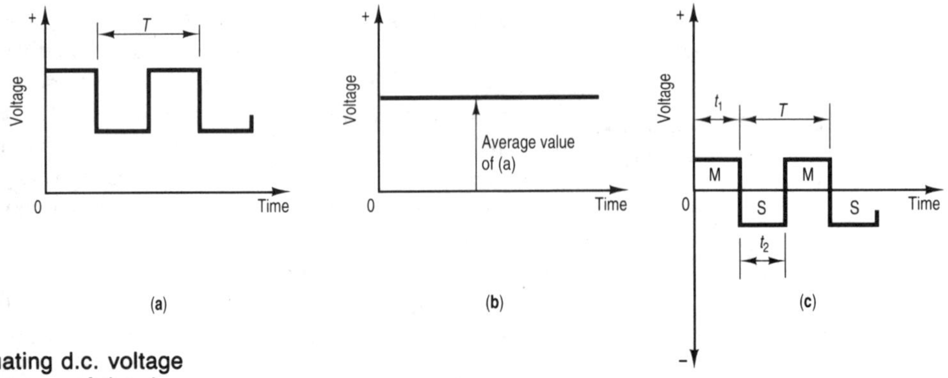

(a)

(b)

(c)

Fig. 6.3 The fluctuating d.c. voltage in (a) is equal to the sum of the d.c. voltage in (b) and the a.c. voltage in (c).

Example 6.1

1. Calculate (a) the mark-to-space ratio, and (b) the duty factor of a 1 MHz rectangular unipolar wave if the positive voltage pulses are 0.2 μs in width.
2. Calculate (a) the mark-to-space ratio and (b) the duty factor of a square wave.

Solution
$T = 1/(1 \times 10^6) = 1 \times 10^{-6} = 1 \ \mu s.$

1. (a) mark-to-space ratio = 0.2/0.8 = 0.25. *(Ans.)*
   (b) duty factor = 0.2/1 = 0.2. *(Ans.)*
2. For a square waveform the mark and space periods are of equal length.

   Therefore:
   (a) mark-to-space ratio = 1. *(Ans.)*
   (b) duty factor = 0.5. *(Ans.)*

Analogue signals

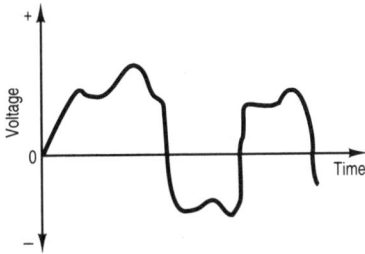

Fig. 6.4 Analogue voltage.

An analogue signal is one whose magnitude varies continuously and may have any value between predetermined limits. The magnitude of the signal changes very little from one instant to the next, as shown by the typical analogue signal given in Fig. 6.4. An analogue signal is very closely related to a physical quantity; the signal shown, for example, could have been produced by a telephone transmitter that has a sound wave incident upon it. Temperature is another analogue parameter that is never able to change instantaneously from one value to another but which always varies gradually. An analogue value, electrical or otherwise, can only be measured approximately even though with expensive instrumentation the degree of accuracy may be very high. A binary digital value, on the other hand, can always be precisely expressed since it is either ON or OFF.

The sinusoidal waveform

The basic analogue signal is the **sinusoidal waveform** shown in Fig. 6.5. The parameters of this wave are:

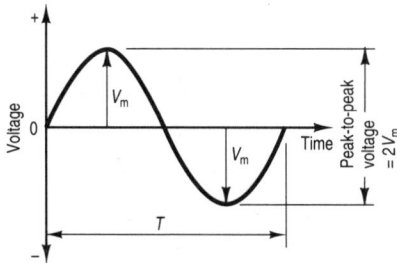

Fig. 6.5 Parameters of a sinusoidal voltage.

1. **Amplitude**. The amplitude, or peak value, of a sinusoidal voltage waveform is its maximum positive or negative voltage measured in volts. This voltage is indicated in the figure by the symbol V_m. The maximum positive and negative voltages are always equal. The **peak-to-peak value** is equal to twice the amplitude, i.e. $2V_m$.
2. **Periodic time**. The periodic time of a sinusoidal waveform is the time, in seconds, taken for the wave to go through one complete cycle of voltage changes. The symbol used to indicate the periodic time is T and this has been shown in the figure.

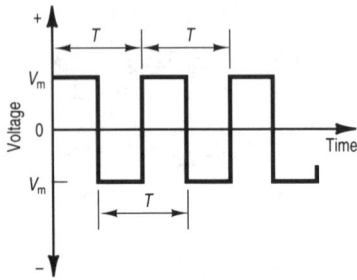

Fig. 6.6 Periodic time of a bipolar rectangular voltage.

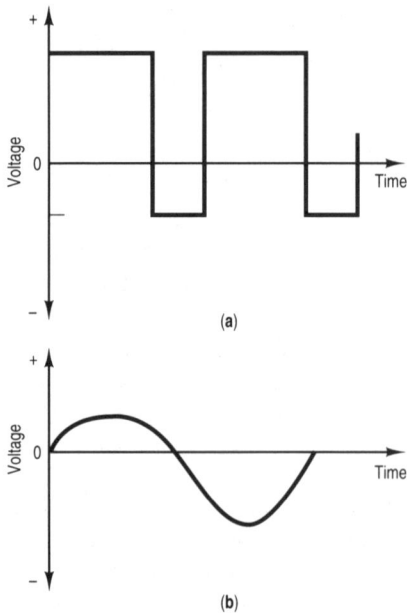

Fig. 6.7 A.C. voltages with different positive and negative peak voltages.

3. **Frequency**. The frequency f of a sinusoidal wave is the number of times per second the wave goes through one complete cycle of voltage changes. The frequency, quoted in a unit known as the Hertz, is equal to the reciprocal of the periodic time. Thus:

$$f = 1/T \text{ Hz} \qquad (6.3)$$

Example 6.2

The frequency of the public mains supply in the UK is 50 Hz. Calculate the periodic time of the supply.

Solution
From equation (6.3), $T = 1/50 = 0.02$ s $= 20$ ms. (*Ans.*)

Non-sinusoidal waveforms

The terms amplitude, frequency and periodic time may also be applied to a non-sinusoidal waveform and Fig. 6.6 shows a *bipolar* rectangular wave whose amplitude varies between equal positive and negative voltages V_m with a frequency of $1/T$ Hz.

The peak positive and negative voltages of an a.c. voltage, or of an a.c. current, are not necessarily equal to one another and Figs 6.7(a) and (b) show two examples of waves with different peak voltages.

Figure 6.8 shows a sinusoidal waveform that has another sine wave, having three times the frequency and 2/5 times the amplitude, superimposed upon it. The lower frequency wave is known as the **fundamental** and the higher frequency wave is known as the **third harmonic** (see p. 000).

Figures 6.9(a) and (b) show, respectively, typical speech and television signal waveforms.

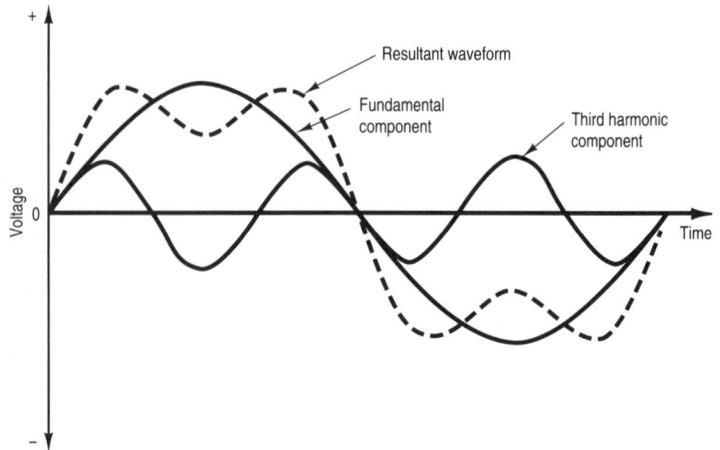

Fig. 6.8 Showing how the sum of a fundamental and its third harmonic produces a non-sinusoidal waveform.

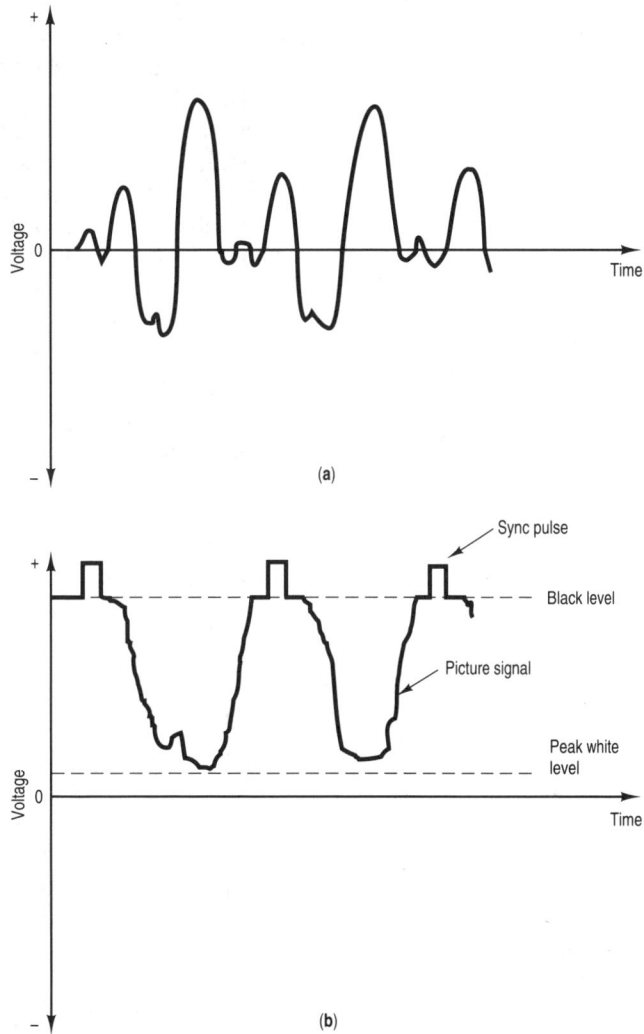

Fig. 6.9 Analogue signals:
(a) speech and (b) television.

Digital signals

A digital signal is one whose amplitude changes abruptly between a number n of defined values. These values can be counted, or measured, exactly whereas the amplitude of an analogue signal can only be approximately known. As a non-electrical example of the difference between analogue and digital values, consider this book: the number of leaves in the book increases in discrete steps of one and is a digital value, but the weight of the book is not known exactly and is an analogue value. Switches on electrical equipment often have three, or more, positions and are digital devices. A common example of this is the waveband switch on a domestic radio receiver, which has long-wave, medium-wave, and FM positions and often also a short-wave position. The voltage of a digital signal changes abruptly from one value to another. Because the signal changes very rapidly between its allowed values a digital signal is said to be **discontinuous**.

Binary digital signals have $n = 2$ and are employed in both digital electronics and computers because **two-state logic devices** can then be used. Such a device has two stable states; the device can either be switched to conduct current, when it is said to be ON, or it can be switched to be non-conducting, when it is said to be OFF. A binary digital signal is one whose amplitude cannot vary continuously but may have only one of two values that are generally known as logic 1 and logic 0 voltage levels. Each binary 1 or binary 0 pulse is known as a binary digit or **bit**. A binary voltage may be either unipolar (i.e. it has only one polarity — positive or negative) or bipolar (i.e. it has both positive and negative values). Within a digital electronic system or circuit unipolar signals are used, but bipolar signals are used for communication between two computers or between a computer and a peripheral. The use of binary digital signals has the advantage that since there are only two possible signal levels they do not have to be known accurately to be correctly recognized by the system.

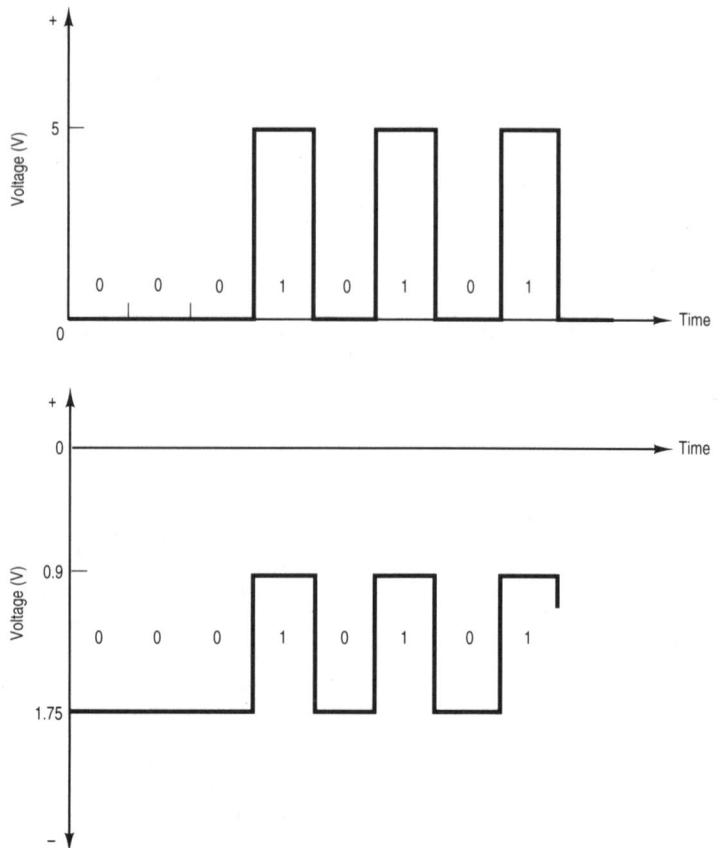

Fig. 6.10 Bipolar binary digital voltages: (a) TTL and (b) ECL.

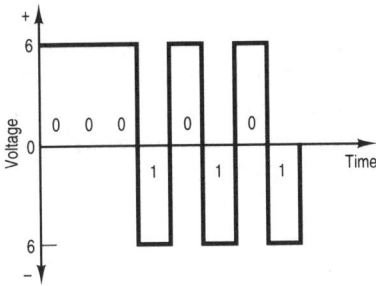

Fig. 6.11 EIA 232 bipolar digital voltage.

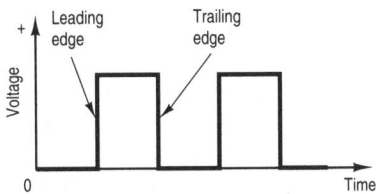

Fig. 6.12 Leading edge and trailing edge of a pulse in a digital waveform.

Fig. 6.13 Decimal number 12 represented (a) by 16-bit pure binary code and (b) by ASCII code.

Depending upon the **logic family** chosen for the digital electronic circuitry employed within a system, the logic 1 voltage level may be high and the logic 0 voltage level low, **or** logic 0 may be high and logic 1 low. The former system is the more commonly employed. Two methods used are illustrated in Fig. 6.10, which quotes typical voltages that are used in the TTL and ECL logic families respectively. For the transmission of digital signals outside of a computer an **interface** standard known as EIA 232 is likely to be used and, with this, interface logic 0 is indicated by a positive voltage in the range 5 to 15 V and logic 1 is indicated by a voltage in the range -5 to -15 V. Often the voltages used are ± 6 V, and this value has been assumed in Fig. 6.11.

Clock waveforms

In a computer, or a microprocessor, or indeed in any other **synchronous** digital circuit or system, the times at which various devices and circuits operate are controlled by a clock waveform. The clock waveform is a unipolar square waveform generated by an oscillator known as the **clock** that has the voltage levels specified by the kind of semiconductor logic in use. The controlled circuits are caused to operate at either the leading edge, or the trailing edge of a clock pulse (these two terms are illustrated in Fig. 6.12).

Example 6.3

The clock in a 20486 PC system operates at a frequency of 25 MHz. Calculate the periodic time of the clock.

Solution
From equation (6.3), $T = 1/(25 \times 10^6) = 40 \times 10^{-9}$ s $= 40$ ns. (*Ans.*)

Signal waveforms

Within a digital system numbers and characters are represented by binary number codes. The most commonly used code is the ASCII code. In the straightforward 16-bit binary code the decimal number 12 is represented by 0000 0000 0000 1100 but in ASCII code 12 is represented by 00110001 00110010. The two binary waveforms that would indicate the number 12 using each code are shown in Fig. 6.13. Clearly the computer must be informed which of the two codes is in use. The waveforms have been drawn with one pulse following another and this is an illustration of **serial transmission**. Serial transmission is a method of moving digital information from one point to another; it is used for communication between two computers, or between a computer and a peripheral that is not located near the computer. The least significant bit of a signal is transmitted first and the most significant bit is transmitted last, or last but one if an error-

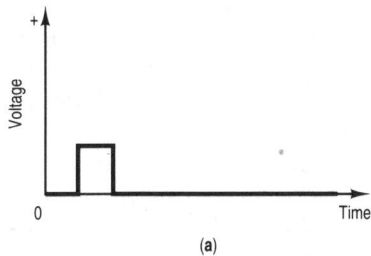

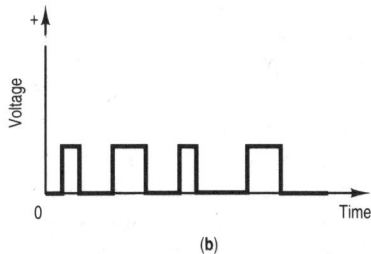

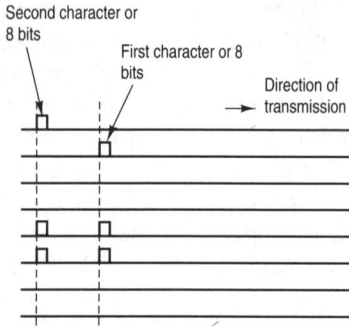

Fig. 6.14 Parallel transmission of data.

Integrated circuits

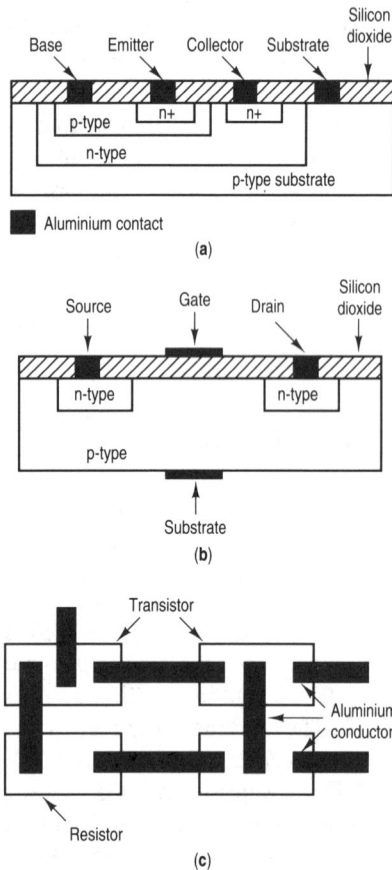

Fig. 6.15 Integrated circuits: (a) bipolar transistor; (b) field-effect transistor; and (c) interconnection of components.

checking bit, known as **parity bit**, is employed.

Within a computer, or microprocessor, information is moved around 8, 16, 32, or even more, bits at a time since this gives very fast transfer of data. Data transfer over a very short computer-to-peripheral connection — for example, between a computer and a disk drive or between a computer and a printer — will also employ **parallel transmission**. Parallel transmission is shown in Fig. 6.14 and it can be seen to employ a separate conductor for the transmission of each bit in a signal.

In modern electronics a circuit or even a complete system can be formed within one small piece of silicon. The various components required, which are mainly transistors with the addition of a few resistors and less often capacitors, are produced by diffusing various other elements into the silicon to produce **p-type regions** and **n-type regions**. A transistor is an electronic device which, in a digital circuit, acts like an electronic switch that is able to switch electronic signals OFF and ON several million times per second with the utmost reliability. Figures 6.15(a) and (b) show how a bipolar transistor and a field effect transistor, respectively, are formed. A resistor consists of a length of n-type silicon. The formed components are then interconnected to produce a particular circuit by forming an aluminium conductive pattern onto the top of the silicon, as shown in Fig. 6.15(c). Each integrated circuit (IC) is then sealed within a suitable package. The majority of IC packages are of the **dual-in-line** type shown in Fig. 6.16. Each IC has two rows of pins to allow the IC to be connected to other ICs, or to other components, to form an electronic system. The number of pins used may vary from 8 to 16 (as shown), up to 48 or sometimes 64.

Most ICs are either linear (analogue) devices or digital circuits. A linear IC is one in which the input and output signals can vary over a continuous range of values. Linear ICs include amplifiers and the various circuits used in radio and television circuits. A digital IC operates with digital input and output signals and examples of such devices include microprocessors and counters.

The internal RAM and ROM of a computer are always provided by ICs. Figure 6.17 shows the pin connections of the Motorola 68020 microprocessor which is used in Apple Macintosh computers. The device has 64 pins, 23 of which are address pins, 16 are data pins, one is a read/write pin, two are power supply voltage pins and two are earth pins. The remainder of the pins have functions that are beyond the scope of this book. The 23 address pins allow $2^{23} = 8388608$ different memory locations to be addressed.

The complexity of a digital IC is loosely defined by placing it into

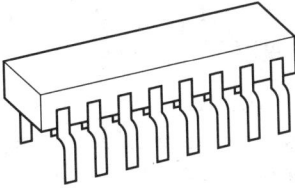

Fig. 6.16 Dual-in-line IC package.

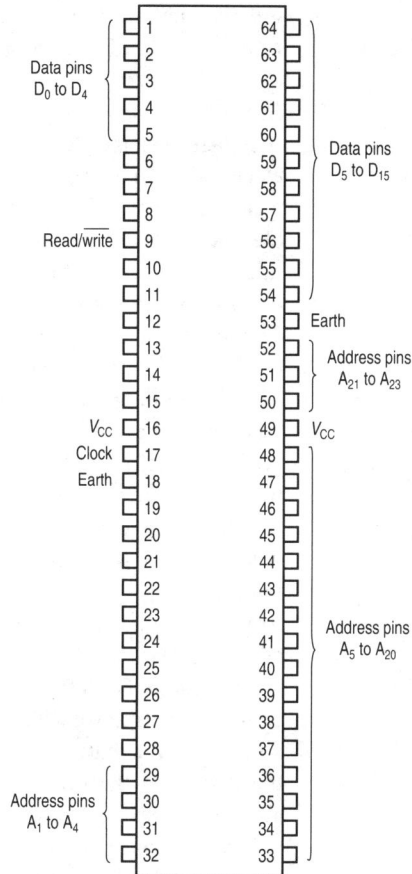

Fig. 6.17 Pin connections of the Motorola 68020 microprocessor.

one of four categories that are based upon the basic digital circuit known as a **gate**:

1. **Small-scale integrated circuit** (SSI). This has no more than 10 gates or equivalent circuits.
2. **Medium-scale integrated circuit** (MSI). This has between 10 and 100 gates or equivalent circuits.
3. **Large-scale integrated circuit** (LSI). An LSI circuit has between 100 and 5000 gates or equivalent circuits.
4. **Very large-scale integrated circuit** (VLSI). A VLSI circuit contains more than 5000 gates or equivalent circuits.

The boundaries between one category and another are not closely defined and they vary in the literature and between different manufacturers.

The use of ICs in a digital system is advantageous because: (a) it allows physically small and lightweight, yet very complex, circuits to be employed; (b) the considerable reduction in soldered connections greatly increase the reliability of a system; and (c) the initial and running costs of a system are both considerably reduced.

Transducers

A **transducer** is a device which converts a non-electrical signal into electrical form or, conversely, converts an electrical signal into a non-electrical form. Transducers that convert from non-electrical into electrical form are also often called **sensors**. A computer or microprocessor is often used to control an industrial process when a sensor of some kind is used to monitor the process to be controlled. The input to a transducer may be a mechanical displacement, a liquid or gas flow, a mechanical force, the level of a liquid, the intensity of light, temperature, or the linear speed or rotational speed of a machine. The electrical signal produced by the sensor may be a voltage, a current, a charge, or a change in resistance. Most transducers are analogue in their operation but a few, like a thermostat, are digital. A computer is only able to handle digital voltages; so, unless a digital transducer is employed, a circuit known as an analogue-to-digital converter (ADC) must be used to convert the signal to digital form (see p. 137). Some examples of simple microprocessor-controlled systems are given in Chapter 7.

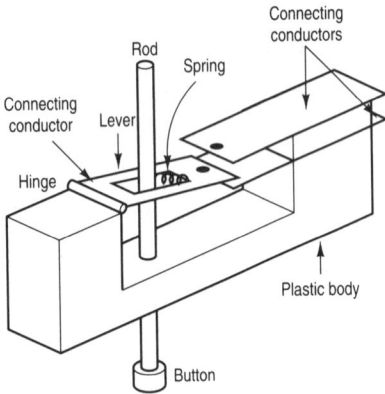

Fig. 6.18 Microswitch.

Any kind of ON/OFF switch is also a digital device and one example of this is the microswitch shown in Fig. 6.18. The switch employs a hinged lever that is able to connect with either of two contacts. In the middle of the lever a slot has been cut through which a rod is able to move perpendicular to the lever when the button at the bottom of the switch is pressed. When the rod is moved a spring is compressed between the rod and the free end of the lever which causes the lever to toggle from one of its contacts to the other.

A thermostat is a digital device because it is either ON or OFF, depending on whether the temperature is above or below a preset value. Figure 6.19 shows the basic construction of a thermostat that is suitable for the control of a domestic central heating system. One end of a bimetallic strip is fixed to the body of the thermostat and the other end, which is free, is located near the button of a microswitch. If the ambient temperature rises, the bimetallic strip bends upwards and pushes against the button to operate the switch. The temperature at which the thermostat operates can be fixed at the required value by means of the adjusting screw.

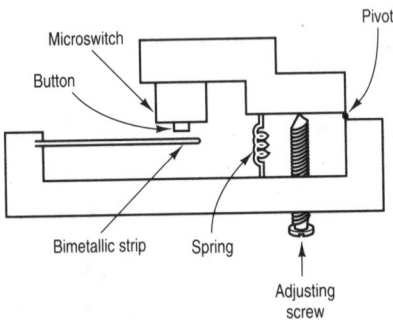

Fig. 6.19 Thermostat.

Temperature

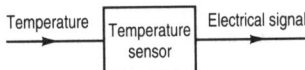

Fig. 6.20 Action of a temperature sensor.

The basic principle of operation of a temperature transducer (or sensor) is shown in Fig. 6.20. The three most commonly employed methods of converting temperature into an electrical signal, when simple ON/OFF operation is insufficient because precise temperature values are required, are: (a) the thermocouple, (b) the thermistor, and (c) the resistance thermometer.

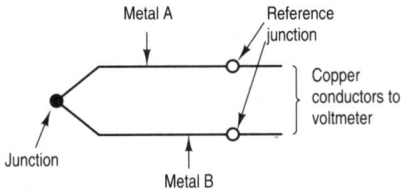

Fig. 6.21 Thermocouple.

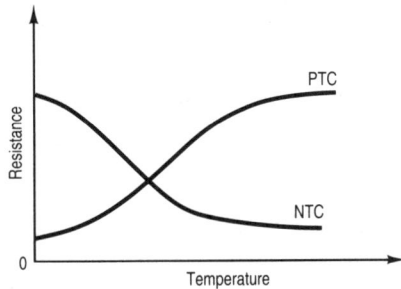

Fig. 6.22 Resistance–temperature characteristics of NTC and PTC thermistors.

The thermocouple

The basic arrangement of a thermocouple is shown in Fig. 6.21. Two conductors, made from dissimilar metals A and B, are joined together at the measuring junction and the other end of each conductor is connected to a sensitive voltmeter. At the junction of the two metals a thermal electromotive force (e.m.f.) is generated which is directly proportional to the temperature of the junction; this voltage is measured by the voltmeter to give an indication of the temperature. The thermocouple is cheap, versatile, and quite reliable and is often used for temperature sensing. For measurement purposes, however, its accuracy is not very good and either a thermistor or a resistance thermometer would probably be used.

The thermistor

A **thermistor** is a device whose electrical resistance varies with temperature. The resistance of the device may either increase or decrease with increase in temperature, depending on the type of thermistor considered. If its resistance increases with increase in temperature the thermistor is known as a **positive temperature coefficient** (PTC) device. Conversely, if its resistance falls as the temperature is increased the device has a negative temperature coefficient of resistance and is known as a **negative temperature coefficient** (NTC) device. The resistance/temperature characteristics of these devices are not linear, as shown by the typical characteristics given in Fig. 6.22. The non-linearity of the devices makes it difficult to use them when accurate temperature measurements are required.

The resistance thermometer

Most resistance thermometers use platinum resistance wire because this material has a linear temperature coefficient of resistance over a wide range of temperatures. This means that it can be used to give an accurate measurement of temperature. When the temperature being measured changes, the resistance of the platinum wire will also change in direct proportion to the change in temperature. The change in resistance is then measured in some way to produce an analogue electrical signal.

Liquid level monitor

The level of a liquid contained in a tank can be monitored and converted into the corresponding electrical signal in several different ways. One possible method is shown in Fig. 6.23(a), which uses a microswitch to indicate whether or not the liquid level has reached some critical point. A float on the surface of the liquid moves upwards with the liquid level, which causes the other end of the pivoted rod

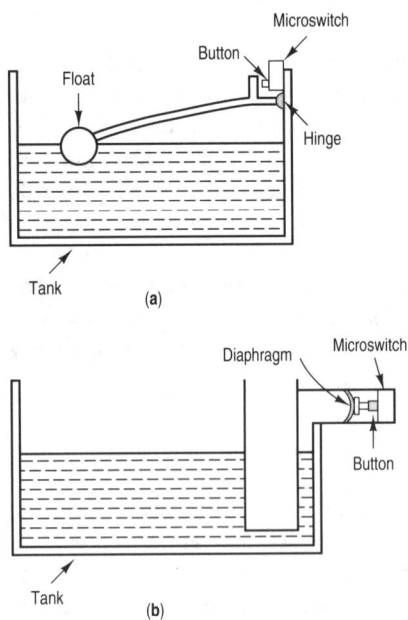

Fig. 6.23 Digital liquid level monitor.

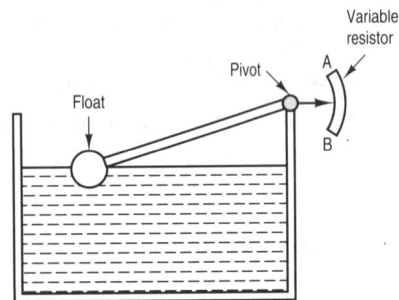

Fig. 6.24 Analogue liquid level monitor.

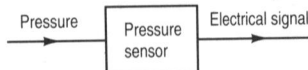

Fig. 6.25 Action of a pressure sensor.

to move towards the button of the switch. When the critical liquid level is reached, the switch button is pressed and the switch is turned ON and in so doing completes the electrical circuit. A current then flows in the circuit to signal that the critical level has been reached. If the float would be an obstacle, an alternative method, shown in Fig. 6.23(b) can be used. As the level of the liquid in the container rises, the air pressure in the pipe also increases and pushes the diaphragm inwards. This, in turn, presses against the button to operate the microswitch.

If the liquid level is to be continuously measured a somewhat more sophisticated method is necessary, and this is shown in Fig. 6.24. Again a float connected to a pivoted rod is used but now the far end of the rod is connected to a wiper that is moved over a variable resistor. When the tank is empty the float is at the bottom of the tank and the wiper is at the point marked A. There is then maximum resistance in the measuring circuit. As the liquid level rises the wiper moves along the resistor towards the point B, and as it does the resistance between the resistor's terminals falls. In this way changes in the level of the liquid contained in the tank are converted into changes in electrical resistance.

Pressure, strain, force, position, and displacement

All kinds of pressure such as water, gas, atmospheric, strain, force, position and displacement can be measured using the appropriate kind of sensor. The transducer is generally either a crystal or resistive strain gauge, or a resistive potentiometer, or a capacitive device. The essential concept of a pressure transducer is illustrated in Fig. 6.25.

Other transducers that are commonly used are the telephone transmitter and the telephone receiver (see Chapter 9); the former device converts sound waves into electrical signals and the latter device converts electrical signals into sound waves. Similarly, the loudspeaker used in both radio and television receivers converts electrical signals into sound waves, the cathode ray tube converts electrical signals into visible light signals. Some transducers — including the electric light bulb, the electric fire, the immersion heater, and the electric motor — are sometimes known as **energy converters**.

Most kinds of transducer give an analogue output electrical signal but a few kinds that are able to generate a digital electrical signal instead are available. One common form of digital transducer is the **shaft encoder**. Essentially, this consists of a coded disk mounted on the end of a rotating shaft, as shown in Fig. 6.26(a), that converts the position of the shaft into a digital code. The disk is made of glass and has the digital code produced by either transparent or opaque sections. A light source is placed to one side of the coded disk and photodetectors are placed on the other side. When the light source is turned on, light is able to pass through the transparent parts of the

Plate with narrow slit

Light source

Contains photo-detectors

Shaft

Circular box

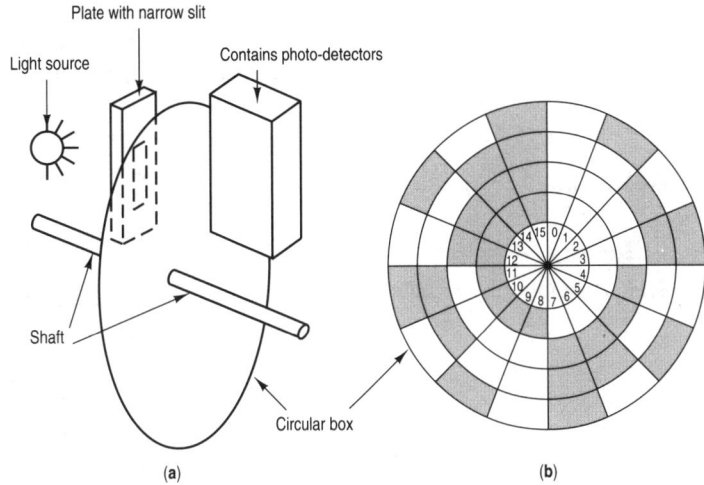

Fig. 6.26 (a) Basic arrangement of a shaft encoder; (b) disk coded using pure binary.

(a)   (b)

disk but not through the opaque parts. The light that passes through the disk is detected by the photodetectors and is converted into the corresponding digital electrical signal. Figure 6.26(b) shows the appearance of a disk that has been coded using pure binary since this is easiest to understand. However, the use of a pure binary-coded disk can lead to large errors occurring and another code, known as the **Gray code**, is normally employed.

Since the position of the shaft may be set by different means, a shaft encoder is not restricted to the conversion of just position information into a digital electrical signal. For example, if the shaft rotates against the known force exerted by a spring, force may be converted, and if the shaft rotation is due to a pressure then pressure may be converted.

Signal conversion

Since most transducers produce an analogue output voltage it is often necessary for the transducer voltage to be converted into digital form before it may be processed or stored by a computer or a microprocessor system. The conversion is carried out by an **analogue-to-digital converter** (ADC). The basic principle of operation of an ADC is shown in Fig. 6.27. The analogue voltage generated by the transducer is applied to the input terminals of the ADC. This voltage is sampled at regular intervals by the ADC and each sample is converted into an n-bit digital signal at the output terminals. The greater the number of bits used to form the output digital word the better will be the resolution of the ADC. If, for example, an 8-bit ADC is considered that has a maximum analogue input voltage of ± 2.5 V then the resolution is $5/2^8 = 5/256 = 19.53$ mV, but for a 10-bit ADC the resolution would be $5/2^{10} = 5/1024 = 4.88$ mV. Clearly, the greater the number of bits used per conversion the better is the resolution but, unfortunately, the more expensive will be the ADC.

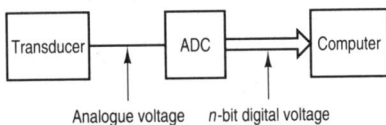

Transducer | ADC | Computer

Analogue voltage n-bit digital voltage

Fig. 6.27 Principle of operation of an ADC.

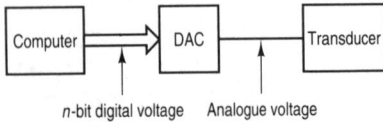

Fig. 6.28 Principle of operation of a DAC.

After the transducer signal has been processed by the computer it is very likely that it will be used to control some physical parameter or process. If the controlled device is analogue in its operation then it will be necessary for the digital output of the computer to be converted into the equivalent analogue voltage. This conversion is the function of a **digital-to-analogue converter** (DAC). The basic concept of a DAC is illustrated in Fig. 6.28. An n-bit digital word is applied to the input terminals of the DAC and the word is then converted into the corresponding analogue voltage. Suppose, for simplicity, that a 3-bit ADC having a maximum output voltage range of 0 to $+5$ V is considered. The output voltage range is divided up into 8 parts ($8 = 2^3$), ranging from 0 to 5 V, so that the resolution is $5/8 = 0.625$ V. When the input digital word is 100 the analogue output voltage will be $4/7 \times 5 = 2.86$ V, and when the digital word is 0010 the analogue output voltage is $2/7 \times 5 = 1.43$ V, and so on for all the other five possible digital words. This means, of course, that the output is not strictly an analogue voltage since it jumps in steps of 0.625 V. If an 8-bit DAC was employed these voltage jumps would each be $5/256 = 19.53$ mV, and for many applications this would present no problem. If it did then a DAC having a better resolution would have to be employed with a consequent extra cost penalty.

When the output of a DAC is used to control an equipment that is not operated by a low voltage it may be necessary to send a current to turn a relay ON and OFF. It is usual to define 4 mA current as representing zero value and 20 mA current as indicating full scale to obtain what is often known as the **4–20 mA current loop**.

7 Microelectronic systems

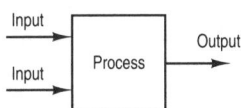

Fig. 7.1 Electronic system.

A system is a connected group of circuits, or of equipments, that is able to perform one, or more, overall functions. The basic block diagram of an electronic system is shown in Fig. 7.1. The input(s) to a system are normally shown as entering from the left with the output(s) drawn as leaving from the right. The system box represents a **process** that is carried out on the input(s) to produce the output(s). The system in the figure has two inputs which are subjected to various processes to produce the single output; but other systems may have different numbers of inputs and perhaps more than one output. Except for the simplest cases a system will consist of a number of subsystems; in a block diagram each subsystem is represented by a separate block. Usually subsystems are separate pieces of equipment that can be individually constructed, purchased, or manufactured. In a domestic central heating system, for example, the subsystems include the gas boiler, the water pump, the control box, the thermostat, and the hot water tank. Examples of electrical systems are many — both large and small.

Three examples of large electrical systems are:

- the **telephone system** which interconnects telephones, cables, transmission systems, and telephone exchanges;
- the **television system** whose subsystems include television studios, television transmitters, television receivers, and both transmitting and receiving aerials;
- the **electricity power supply system** that supplies electrical power to the home, the office, and the factory.

Small electrical systems include:

1. **Domestic radio receiver**. The inputs to a domestic radio receiver are the radio signals picked up by the aerial and the electrical power taken either from the mains supply or from a battery. The basic block diagram of a radio receiver is shown in Fig. 7.2. The output from a radio receiver is the sound, music or speech produced by the loudspeaker. The processes carried out by the radio receiver system include the selection of the wanted signal from all those signals that are picked up by the aerial, the shifting

Aerial

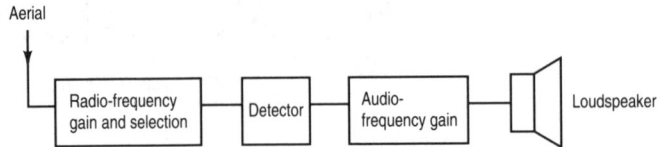

Fig. 7.2 Radio receiver.

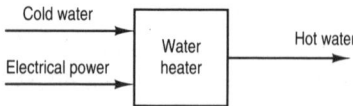

Fig. 7.3 Water heater.

of the wanted signal from its original radio frequency to the audio bandwidth, and the amplification of the selected signal so that it has enough power to operate the loudspeaker.

2. **Water heater**. The inputs to a water heater are the electrical power supplied by the mains electricity and the cold water, while the output is heat energy and hot water. The process carried out in this case is the conversion of electrical energy into heat energy which, in turn, raises the temperature of the water to the required value. The block diagram of a water heater is shown in Fig. 7.3.

3. **Video recorder**. The inputs to a video recorder are the television signals that enter at the aerial socket, the programme commands that are entered by the user, and the electrical power supply. The entered commands instruct the recorder when to record a certain programme and for which period of time. Several such commands may be entered at the same time, the maximum number depending upon the particular model concerned, and the recorder will obey each command in chronological order. The output from the recorder is the wanted programme(s) recorded on video tape.

The three examples of small systems are not all examples of **electronic systems**, however. A water heater contains no electronics at all and so it is not an electronic system. Other systems may include some electronic circuitry but still would not be classed as electronic systems; one obvious example of this is the modern car, which contains quite a lot of electronic circuitry but is clearly a mechanical system.

An electrical system can be used without any understanding of the way in which the system works. The user of a television receiver needs only to know how to turn the receiver on and how to select different channels; fortunately, no knowledge of how the receiver converts the incoming television signals into a visible picture is required. Similarly, the user of a telephone needs only to dial, or key, the wanted number and requires no knowledge of how each call is routed to its correct destination.

Use of microelectronic devices

In recent years the cost of electronic control has fallen dramatically largely because of the widespread use of microelectronic devices such as the **microprocessor**. For some time now electronic control has been employed in domestic appliances such as washing machines, microwave ovens, and video recorders, but now it is also being increasingly used for more mundane applications such as coffee makers, irons, and vacuum cleaners.

Microprocessor applications can be broadly divided into two main groups. Firstly, there is the use of a microprocessor in a small computer where it can be repeatedly re-programmed to perform different tasks. Every time a new program is run by the computer the processor handles a different set of instructions. Secondly, there are the microprocessor applications in which the device is programmed once only and it is then used to execute the same program over and over again. The program is generally held in a ROM, i.e. it is firmware. This method of operation is known as **embedded control** and is employed for the control of household appliances and central heating systems, for car electronics, and for the control of many industrial equipments and machinery. Thus, embedded control is a microprocessor application in which the microprocessor is programmed once only at the factory.

The microelectronic device used to control a domestic appliance must be cheap, reliable, and safe and some processors have been specifically designed for such purposes. The processor used in a household appliance must, for example, contain circuitry to ensure that a heater cannot be left on for a long period of time to avoid the obvious fire risk. A potentially common cause of processor failure is **brown-out** of the mains power supply. Brown-out is said to have occurred whenever the supply voltage to the processor has fallen below a minimum permissible value and so corrupting the contents of both RAM and registers. The brown-out condition may put a processor into a potentially dangerous state and hence additional brown-out detection circuitry is provided. This circuitry resets the processor if a brown-out should take place.

Microprocessors, microcomputers, and microcontrollers

The processor used in a microelectronic system may be a microprocessor, a microcomputer, or a microcontroller. A **microprocessor** is an IC that contains just the circuits, such as the arithmetic and logic unit (ALU), and accumulator and registers that form the central processing unit (CPU). The chip does not contain a RAM or a ROM (in some cases there may be a small amount) or input/output interface circuitry. The RAM and/or ROM that is needed must be provided externally along with the necessary input/output interface circuits. A microprocessor is used, for example, in smaller computers such as the Intel 20386 and 20486 in PCs, the Motorola 68020/30/40 in Apple Macintosh machines, and the Zilog Z80 in Amstrad PCW machines. The 68000 family devices are also commonly employed for industrial control applications.

A **microcomputer** IC includes on-chip memory, RAM and/or ROM, and input/output circuitry as well as a CPU. In other words it is a complete computer on-a-chip. Generally, a microcomputer is used in small computer-controlled systems which do not require many different computing facilities. A microcomputer is able to run a variety of programs that are loaded in from disk.

A **microcontroller** is essentially a microcomputer that has been specifically customized for use in a domestic or industrial control application. Because simpler applications require simpler programs the CPU usually needs fewer registers, only one accumulator, and a minimum of both RAM and ROM. It will also have all the necessary input/output interface circuitry on-chip. This means that a micro-controller is a simpler, less powerful, version of a microcomputer. There are a large number of different microcontrollers on the market; many of them use the same CPU but are provided with different on-chip facilities such as an assembler or a C compiler, input/output interfaces, and memory.

The performance of a microcontroller depends to a large extent upon the width of its internal data bus. Bus widths of 4, 8, 16, and 32 bits are employed in different devices but the higher the number of bits used the more expensive is the device. A 4-bit controller is used in calculators, and for simple systems in which only small amounts of data need to be processed using a simple program and the operation time is not important. An 8-bit controller gives a greater speed of operation and is used for more demanding applications such as many automobile electronic systems and washing machines. 16-bit and 32-bit controllers are also available and are used in complex applications such as video recorders, laser printers, disk drives, and signal processors.

Microprocessor systems

The basic block diagram of a microprocessor system is shown in Fig. 7.4. Essentially it consists of the microprocessor IC itself, a crystal oscillator that acts as the system clock, RAM and ROM, and an input/output interface IC.

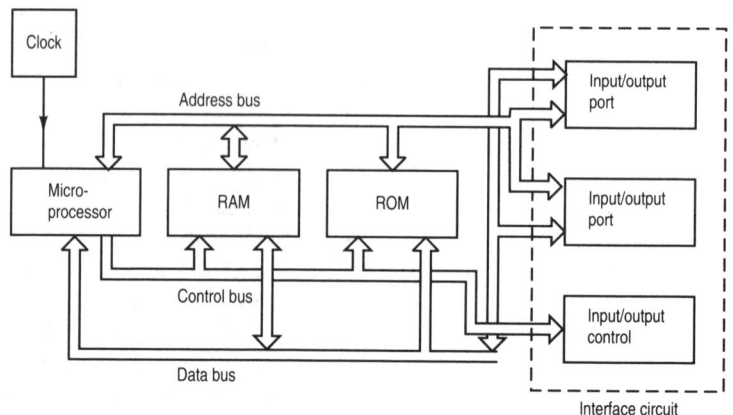

Fig. 7.4 Basic microprocessor system.

Clock

The instructions contained within a program must be carried out in the correct sequence and all parts of the microprocessor system must operate at exactly the correct instants in time, i.e. they must be in synchronization. The clock waveform is applied to a timing and control unit inside the microprocessor and this unit generates signals that operate the various circuits in the system at the correct instants in time. These control signals include $\overline{CS}$ that makes a memory IC active, and read/$\overline{write}$ that puts the IC into either its read or its write mode.

Microprocessor

The internal architecture of a microprocessor IC is complex and varies from one device to another. All microprocessors, however, have certain features in common. They all contain a number of registers which are used for the temporary storage of data, an arithmetic and logic unit (ALU) that is used to carry out arithmetic and logic instructions, and timing and control logic. Microprocessor ICs may be either 8-bit, or 16-bit, or 32-bit devices. The earliest devices were all 8-bit and the more commonly employed included the Intel 8080 and 8085, the Zilog Z80, the Motorola 6800/6809, and the MOS Technology/Rockwell 6502. Large numbers of these microprocessors were used in all kinds of equipment including microcomputers, and for many control applications they are still often used. Later developments have led to the introduction, first, of 16-bit and, later, of 32-bit microprocessor ICs. The increased number of bits gives faster operation, a wider instruction set (including multiply and divide), more addressing modes, and a larger addressable memory. 16-bit devices include the Intel 8086, 8088, 80186 and 80286, the Motorola 68000 family, and the Zilog Z8000 family. The latest 32-bit microprocessors include the Intel 80386, 80486 and Pentium, the Motorola 68020 and the Zilog Z80000. The later introductions are usually software compatible with their predecessors; thus the 80386/80486 devices are compatible with the 80286 and the even earlier 8088, and the 68020 is compatible with the 68000.

RAM and ROM

RAM and ROM are memory devices that provide longer term storage of data than do the internal registers. Data can either be written into or read out of a RAM and data is held in a particular location until either new data is written in or the power supply is removed. The data held permanently in a ROM can only be read out; it cannot be altered in any way.

Input and output circuits

A port is an interface circuit that provides the necessary logic and buffering. An input port is used to connect signals from peripherals to the microprocessor. An output port allows the microprocessor to send signals to a peripheral. Peripherals include keyboards, monitors, printers, and disk drives. An interface IC has two or more ports that may be programmed to act as either an input or an output. Some devices allow the lines within a port to be individually programmed as IN or OUT lines although even then usually all lines in a port are similarly programmed. To input or output data the microprocessor places the address of the port onto the address bus. Then, if input data is to be read the microprocessor puts an input/output read signal onto the control bus. This signal causes the interface circuit to place the contents of the input buffer of the addressed port onto the data bus and this data can then be loaded into the accumulator. If data is to be outputted the microprocessor places the address of the port onto the address bus, places the data onto the data bus, and puts an input/output write signal onto the control bus. The interface circuit then moves the data from the data bus and into the output buffer of the addressed port.

The basic arrangement of a **peripheral interface device** is shown in Fig. 7.5. The interface device is a LSI circuit which is a member of the same family as the microprocessor; some circuits are made for use with a parallel interface and others are made for use with a serial interface. A serial interface IC must also provide both parallel-to-serial and serial-to-parallel conversion of data. Examples of each type of interface IC can be seen in Figs 7.6(a), (b), and (c). In parts (a) and (b) a parallel interface connects a keyboard or an ADC to the microelectronic system via port A, which has all its lines configured to act as inputs, while part (c) shows a serial interface to a **modem** (p. 209).

Different names are used by different manufacturers for these interface devices; they are called variously a **parallel input/output (PIO)**, a **universal synchronous/asynchronous receiver/transmitter (USART)**, an **asynchronous communication interface adaptor (ACIA)**, a **versatile interface adaptor** (VIA), and so on.

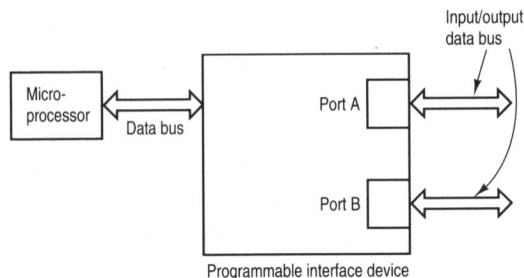

Fig. 7.5 Programmable interface device.

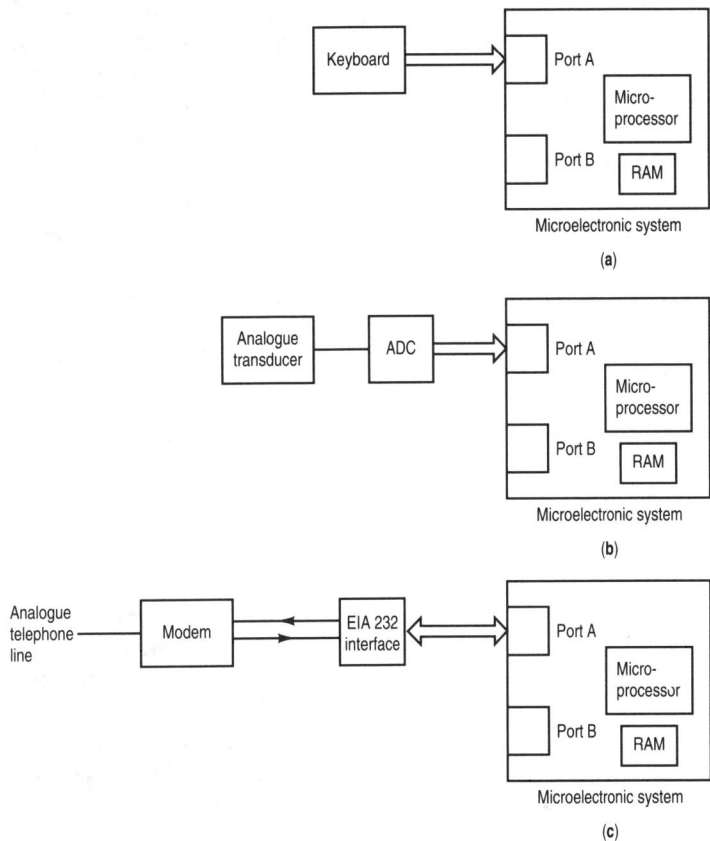

Fig. 7.6 Use of a programmable interface device: (a) and (b) parallel interfaces, (c) serial interface.

A memory-mapped input/output interface IC is, as far as the microprocessor is concerned, identical to a memory location. To output data from a port the microprocessor addresses the port in exactly the same way as it addresses a memory location before storing data in memory. Some microprocessors use a different technique; separate IN and OUT commands are used to send or receive data to or from a port.

Bus lines

The connections between the various circuits in a microprocessor, and between a microprocessor and both the internal memory and the input/output interface ICs, are provided by **buses**. There are three buses: the address bus, the control bus, and the data bus. Each bus consists of a group of conductors giving a width of 8, 16, or 32 bits. The **control bus** carries control signals, such as $\overline{CS}$ and read/write, from the timing and control unit to the other parts of the microprocessor. The control bus is unidirectional. The **data bus** carries machine instructions and data between the memory and the microprocessor and data to and from the input/output interface IC. The data bus is bidirectional — which means that the microprocessor can write data onto the bus to be written into a memory location *or* it can read data from a memory location. The **address bus** carries the address of a location in memory, or of an input/output port, that is to be used in a transfer of data. The address bus is also unidirectional.

At present, 8-bit microprocessors are still being sold by semiconductor distributors but they are rarely, if ever, used for new designs because of their limitations. These limitations include the following:

• Data words longer than 8 bits can only be operated on by splitting the word into two parts and dealing with each part separately. This process considerably increases the time taken to execute an instruction.
• Many applications require more memory space than an 8-bit microprocessor is able to access directly.
• Most 8-bit microprocessors are unable to perform multiplication and division directly but must make use of a subroutine.

The 16-bit microprocessor has been designed to give an increased speed of operation. Besides being able to move data around 16 bits at a time a **pipelining** technique is often used that allows an instruction to be fetched from memory before the previous instruction has been completely executed. Cache memory (p. 37) may also be employed to increase the speed still further.

Operation of a microprocessor

The simplified block diagram of a microprocessor is shown in Fig. 7.7. The labelling of the index registers will vary from one kind of microprocessor to another and some devices will contain several registers. Suppose that this basic microprocessor is to be used to add together two numbers, say 6 and 3, that are held in memory locations 0080 and 0081 respectively. The SUM of the two numbers is to be stored at memory location 0082. The first instruction in the program

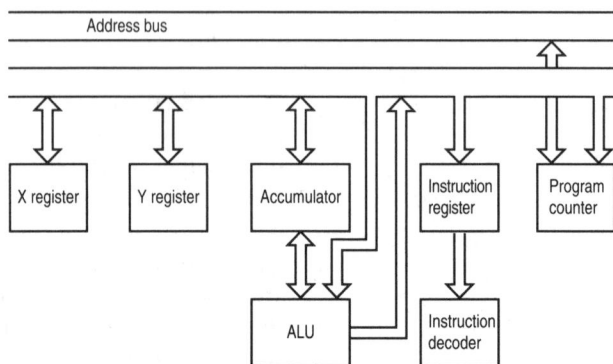

Fig. 7.7 Basic internal block diagram of a microprocessor.

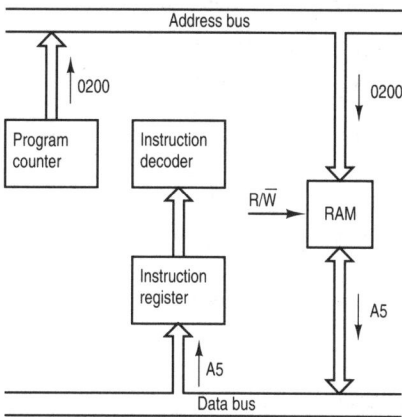

Fig. 7.8 Operation of a microprocessor: step 1.

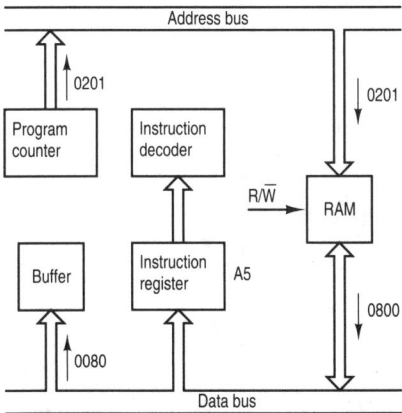

Fig. 7.9 Operation of a microprocessor: step 2.

is held at memory location 0200. All these numbers are quoted using hexadecimal notation.

The action of the microprocessor in carrying out this relatively simple operation is best considered in a number of steps.

Step 1. The first step is for the microprocessor to address the memory location that holds the first program instruction and then to fetch that instruction. This initial step is shown in Fig. 7.8. The address held in the program counter, i.e. 0200, is sent, via the address bus, to the RAM. The read/write (R/$\overline{\text{W}}$) line is then taken high by the instruction decoder and the instruction held at location 0200 is read out onto the data bus. This instruction will be the machine code which indicates the operation LOAD Accumulator; here it is assumed that this machine code is A5. The A5 instruction is held by the instruction register and decoded by the instruction decoder, and the program counter is incremented to hold the address of the next instruction, i.e. 0201.

Step 2. It is now necessary for the microprocessor to fetch the next instruction that gives the address of the memory location storing the first of the two numbers to be added. Figure 7.9 shows the position; the program counter holds 0201 and the data read from memory and loaded into a buffer is 0080.

Step 3. This step is illustrated in Fig. 7.10. The address, 0800, of the memory location that holds the first number is transferred from the data buffer into the program counter. This memory location is addressed and then the R/$\overline{\text{W}}$ line goes high and the number, 6, stored at the location is read out and placed onto the data bus. This number is then moved into the accumulator.

Step 4. The contents of the program counter are then changed to 0202 to point to the address of the next program instruction. Memory location 0202 is then accessed and the R/$\overline{\text{W}}$ line is taken high so that the contents of the addressed location can be read. The next instruction is therefore fetched via the data bus to the instruction register. It is assumed that the machine code instruction for ADD with

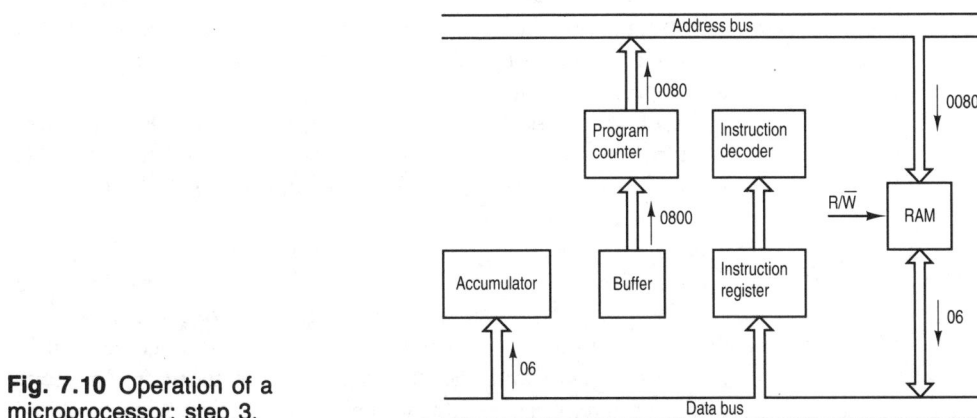

Fig. 7.10 Operation of a microprocessor: step 3.

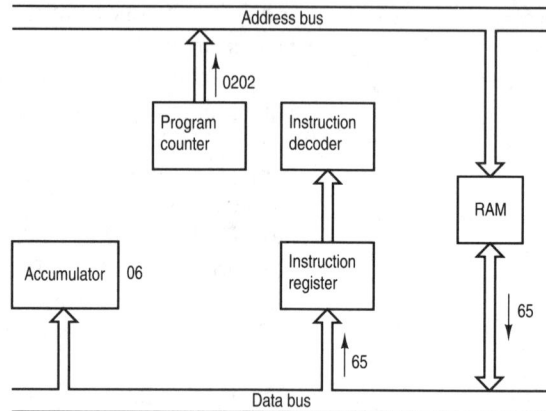

Fig. 7.11 Operation of a microprocessor: step 4.

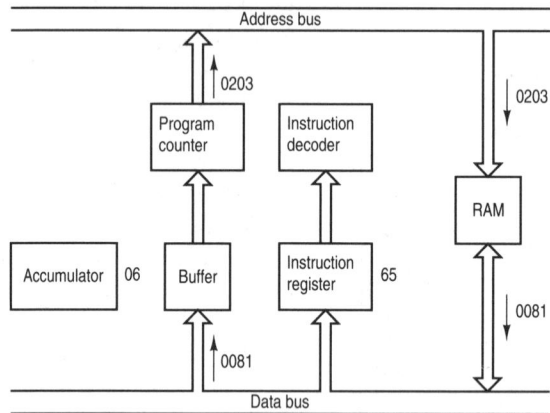

Fig. 7.12 Operation of a microprocessor: step 5.

carry is 65. This code is recognized by the instruction decoder which then waits for the second number to arrive. Figure 7.11 shows the situation after the fourth step in the process.

Step 5. The contents of the program counter are now incremented to 0203 as shown in Fig. 7.12, and the contents of this memory location are fetched from memory. This time the data fetched is the address, i.e. 0081, of the memory location where the second number is stored. This data is placed into the buffer and is then transferred to the program counter.

Step 6. Now, as shown in Fig. 7.13, the memory location 0081 is addressed and the number, 03, held there is read out onto the data bus and then into the data buffer. The number is held in the buffer while the addition process is carried out.

Step 7. The next step in the process is shown in Fig. 7.14. The program counter contains the address of the next instruction to be executed, the ADD with carry instruction is held in the instruction decoder, and one of the two numbers to be added together is held in the accumulator while the other is held in the data buffer. On receipt of a command from the instruction decoder the two numbers are placed

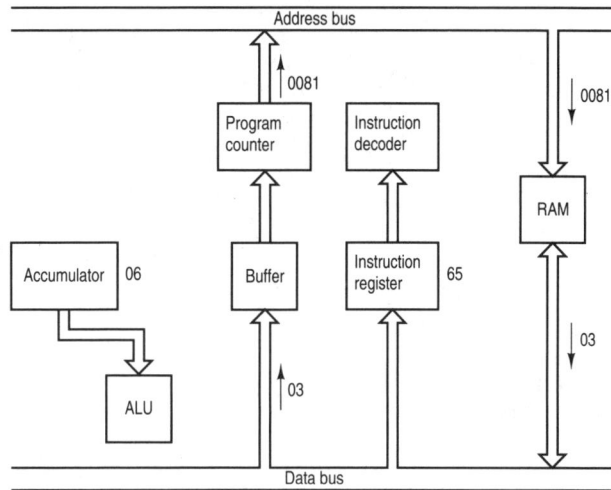

Fig. 7.13 Operation of a microprocessor: step 6.

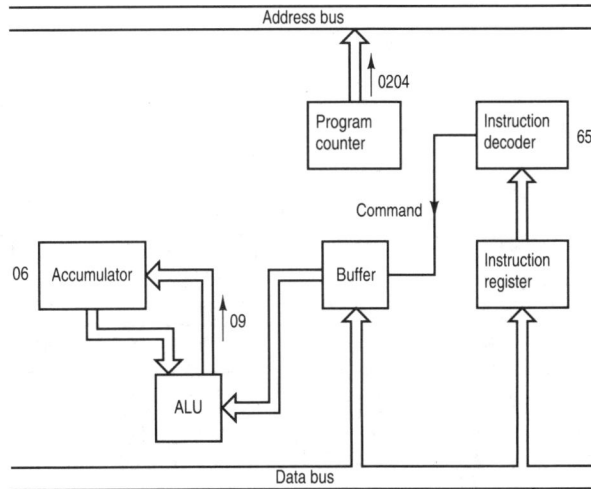

Fig. 7.14 Operation of a microprocessor: step 7.

into the ALU where their addition is performed. The SUM that is produced is then placed into the accumulator.

Step 8. It is now necessary for the SUM to be stored in memory at location 0082. The next instruction is pointed to by the program counter and is stored at location 0204. When the R/$\overline{\text{W}}$ line is taken high the contents of this location are read out and are placed in the instruction register. It has been assumed that the machine code for STORE Accumulator is 85. The contents of the relevant parts of the circuit are shown in Fig. 7.15.

Step 9. Now the program counter is incremented so that it holds 0205 and the contents of this memory location are read out, placed on the data bus, to appear in the data bus buffer. The contents are the address of the memory location in which the SUM of the addition is to be stored, i.e. location 0082. This is shown in Fig. 7.16. The

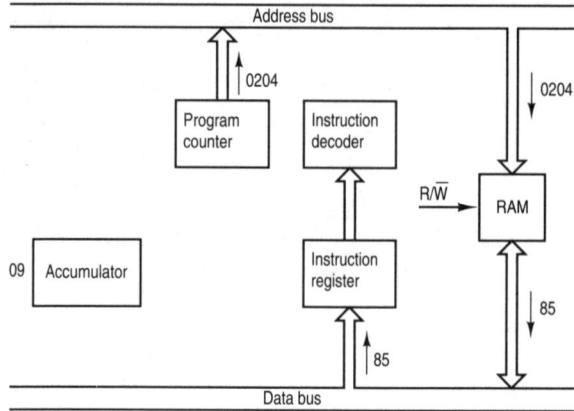

Fig. 7.15 Operation of a microprocessor: step 8.

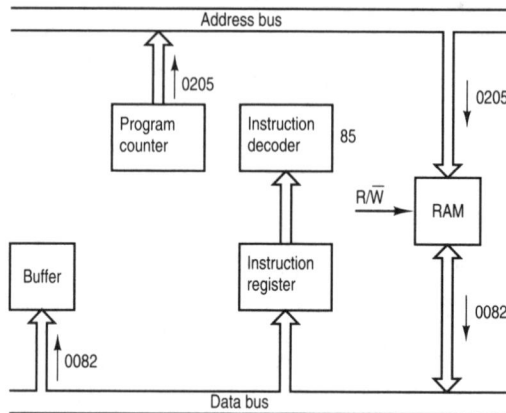

Fig. 7.16 Operation of a microprocessor: step 9.

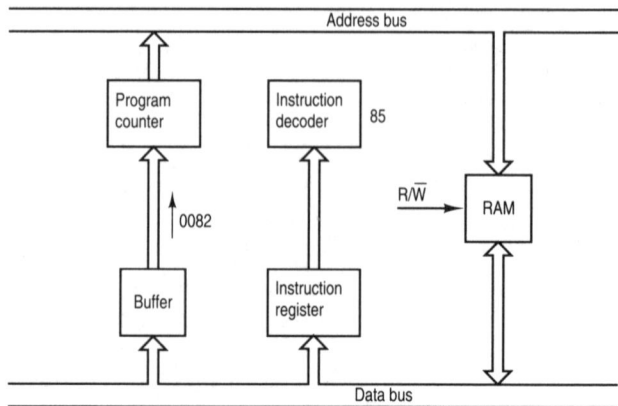

Fig. 7.17 Operation of a microprocessor: step 9.

contents of the buffer are then transferred into the program counter (see Fig. 7.17).

Step 10. The last step in the process is to transfer the contents of the accumulator, i.e. the SUM, to location 0082 in the memory. The way in which this is done is shown in Fig. 7.18. The contents of the

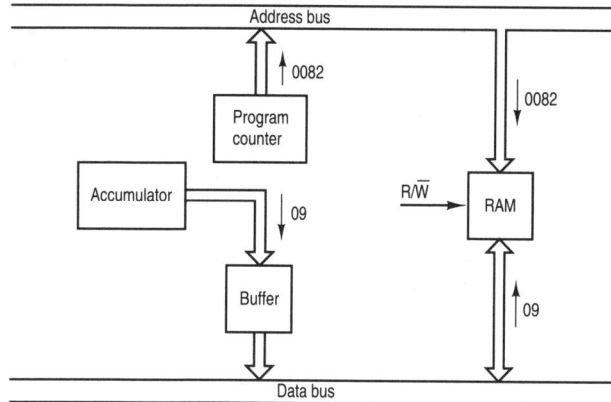

Fig. 7.18 Operation of a microprocessor: step 10.

accumulator are transferred to the data buffer and then, via the data bus, to the memory. Once the location 0082 has been accessed and the data is available at the memory, the R/$\overline{W}$ line is taken low and the data is written into that location.

[The process described above has been simplified by using the simplest form of memory addressing — that is, by not showing how the program counter is supplied with the next instruction address after it has had an address transferred to it from the data bus buffer, by not showing how the read/$\overline{write}$ signals are obtained, and so on.]

Microprocessor-controlled systems

A microprocessor, or a microcontroller, can be used to control the operation of a wide variety of machines and equipments, ranging from domestic appliances to industrial processes.

Automatic washing machine

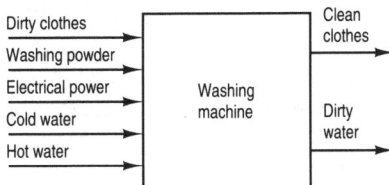

Fig. 7.19 System diagram of an automatic washing machine.

The system block diagram of an automatic washing machine is shown in Fig. 7.19. The inputs to the system are hot and cold water, dirty clothes, washing powder and, of course, electrical power.

The block diagram of an automatic washing machine is shown in Fig. 7.20. The functions of the blocks shown are as follows:

1. **Cold-water inlet valve**. The cold-water inlet valve can be turned on or off to allow cold water into the machine.
2. **Hot-water inlet valve**. The hot-water inlet valve controls the flow of hot water into the machine.
3. **Dirty water output**. The pump is switched ON to pump dirty water out of the machine.
4. **Heater**. The electric heater heats the hot water taken into the machine up to the required temperature.
5. **Water level sensor**. The function of the water-level detector is

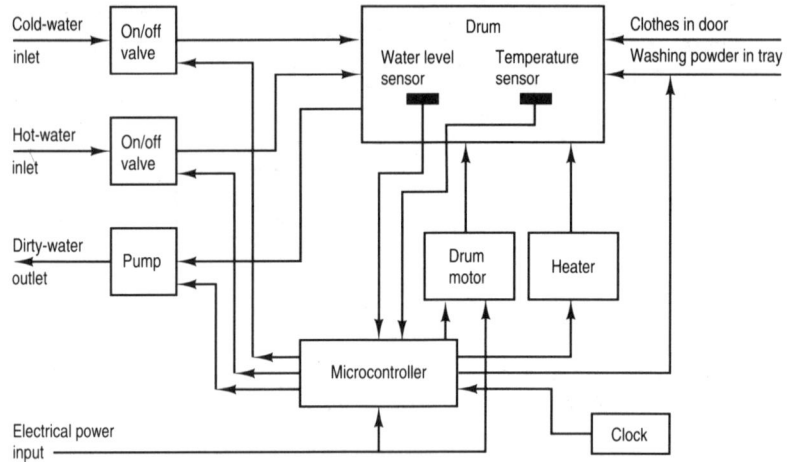

Fig. 7.20 Block diagram of an automatic washing machine.

to indicate when the water level inside the drum has reached the required depth.

6. **Temperature sensor**. The temperature sensor signals to the microcontroller the temperature of the water inside the drum.

7. **Drum motor**. An electrical motor is used to rotate the wash drum at any required speed; the speeds vary according to the part of the wash, or rinse, or spin cycle reached by the machine.

8. **The microcontroller**. The microcontroller reads the wash program set by the user of the machine and chooses the appropriate built-in program (in ROM) to follow. The controller then controls the actions of all the other parts of the machine in accordance with the chosen program.

9. **The clock**. The clock controls the time for which each process in the washing machine cycle lasts.

The sequence of operations carried out by the washing machine is controlled by the selected program that is permanently held in the ROM. The controls of the machine allow the user to select any one of the built-in sets of washing sequences. The user places the dirty washing into the machine and washing powder into a container. A washing program is selected and then the 'start' button is pressed. The basic washing sequence followed by the machine is given below.

1. The controller reads the control settings.

2. The machine is filled with hot water taken from the domestic supply. The inflow of hot water ceases when the level detector senses that the correct level has been reached and signals that fact to the controller. The controller then closes the hot-water inlet valve.

3. The hot water in the drum is then heated to the temperature required by the chosen wash cycle. The heater is turned off when the temperature sensor indicates to the controller that the correct temperature has been reached.

4. The clothes in the machine are then washed for a set period of

Start

Read set program

Fill with hot water

Heat water

At required temperature? — No

Yes

Input washing powder

Turn drum at wash speed

Set time up? — No

Yes

Stop drum & pump out dirty water

Fill with cold water

Turn drum at spin speed(s)

Set time up? — No

Yes

Stop

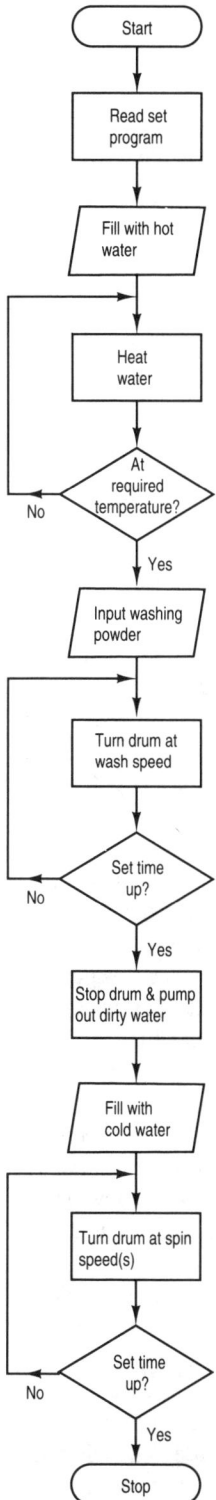

Fig. 7.21 Flow chart of the operation of an automatic washing machine.

time that is determined by the chosen wash cycle. The washing process involves the rotation of the drum at fixed speeds, and in opposite directions, and the taking in of washing powder and perhaps conditioner.

5. At the end of the washing period the dirty water is pumped out of the drum.
6. The cold-water inlet valve is then opened to allow cold water to enter the drum. The inflow of cold water stops when the level detector signals to the controller that the correct water level has been reached and then the controller closes the cold-water inlet valve.
7. The clothes are rinsed for a set length of time by rotating the drum at a fixed speed.
8. The cold water is pumped out of the drum.
9. The clothes are now spun dried for a set time; many machines go through a sequence of different drum speeds to obtain the optimum spin-dry action.
10. At the end of the spin-dry action the machine is automatically turned off. An audible alarm may be given to alert the user that the wash has been completed.

The flow chart for the operation of an automatic washing machine is given in Fig. 7.21.

Central-heating system

A gas-fired central-heating installation is another example of a system. The inputs to the system are the gas supply, the electric power supply, cold water, and the commands entered by the user. The processes carried out are the heating of the water to heat the radiators and the heating of the hot water supply.

A domestic gas-fired central-heating system is used to maintain the temperature inside a house to within a small range of values either side of the value set on a control box. The air temperature is monitored by a temperature sensor. The controller has two inputs: (a) the measured temperature and (b) the temperature set on the control box by the user. The temperature of the house is controlled by comparing the actual room temperature with the temperature set by the user. Whenever there is a difference between these two values the system will operate to reduce the difference, nominally to zero. The system will also maintain the temperature of the hot water supply at, or near, another set temperature.

In a non-microprocessor-controlled system the temperature sensors are on/off thermostats. The basic arrangement of a domestic central-heating system is shown in Fig. 7.22. The hot water produced by the gas boiler is circulated around the pipes and radiators in the system by an electrically operated water pump. The pump is switched on and off by the air temperature thermostat. When the pump is operating

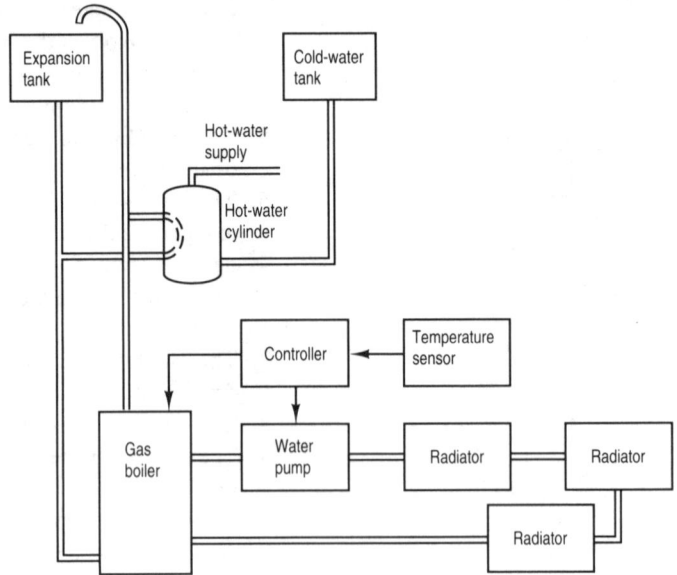

Fig. 7.22 Central-heating system.

hot water is forced around the system, heating up the radiators and hence the rooms in which they are fitted. When the thermostat senses that the wanted room temperature has been reached it operates and turns the water pump off. A thermostat always exhibits **hysteresis**; this term means that the thermostat switches off at a slightly higher temperature than that at which it switches on again.

Hot water from the boiler also circulates by natural gravitation (i.e. heat rises) through a pipe coil inside the hot-water cylinder. Here heat is exchanged between the coil and the water inside the cylinder to raise its temperature. When the water temperature has risen to the required value the tank thermostat operates and causes a water valve to close so that no more hot water passes into the hot-water tank. Lastly, the gas boiler heats up the water passing through it to a temperature set by another thermostat mounted inside the boiler.

Because a thermostat exhibits hysteresis neither the room nor the water temperatures are maintained constant at exactly their set values. Instead the temperature of each varies in a more or less sinusoidal manner either side of the set value. A typical graph showing how the temperature of a thermostatically controlled heating system varies with time is given in Fig. 7.23.

More precise control of the temperature is possible if the central heating system is microprocessor controlled. The microcontroller can take over the function of the system controller and control the times for which the heating and/or the hot water are turned on and off. Instead of a simple on/off temperature sensor like a thermostat an analogue temperature transducer must be employed. The input ports

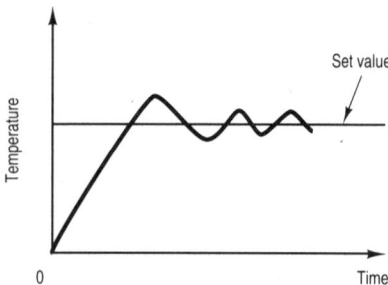

Fig. 7.23 Variation of temperature with time of central heating.

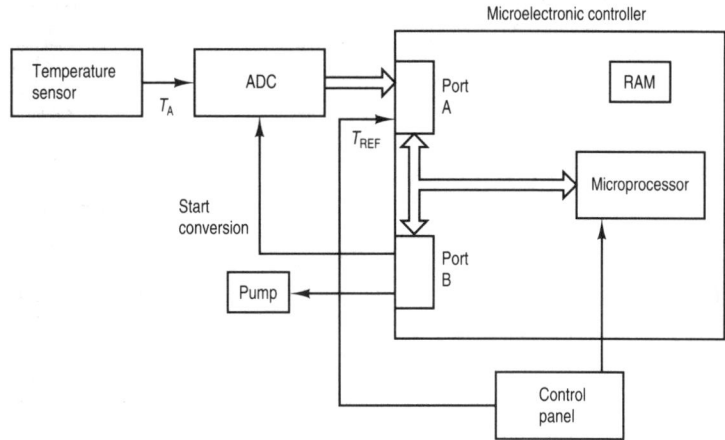

Fig. 7.24 Microprocessor control of a central-heating system.

of the microcontroller would be connected to the room and the hot-water cylinder temperature sensors. The output ports would be connected to the boiler and to the water pump to allow them to be switched on and off. The basic principle of microprocessor control of a central-heating system is shown in Fig. 7.24. The required temperature T_{REF} is entered at the control panel and is stored in a RAM. It may sometimes be possible to also enter a required tolerance on T_{REF}; if so this tolerance is also entered in a RAM. The control program will be stored in a ROM.

The ambient temperature T_A at the location of the temperature sensor is monitored to produce an analogue voltage that is proportional to temperature. This analogue signal is then converted into digital form by the ADC. The digital word representing the ambient temperature T_A is compared with the reference temperature T_{REF}. If $T_A < T_{REF}$ a control signal, probably binary 1, is sent via port B to the water pump switch to turn the pump on; hot water then circulates around the system and the air temperature rises. Conversely, if $T_A > T_{REF}$ another control signal, probably binary 0, will be sent to turn the pump off. If tolerances $\pm \Delta T$ about T_{REF} have been entered at the control panel then the switching action will take place when either $T_A > (T_{REF} + \Delta T)$ or $T_A < (T_{REF} - \Delta T)$. The flow chart describing this action is given in Fig. 7.25

The basic heating system of Fig. 7.24 can be extended in several different ways. The microprocessor can be programmed, at the control panel, to turn the heating on or off at set times of day, and the system can also control the hot-water supply and the operation of the gas boiler. In addition, the control panel may display the time, the status of the system, e.g. heating on or off, the ambient temperature, and so on.

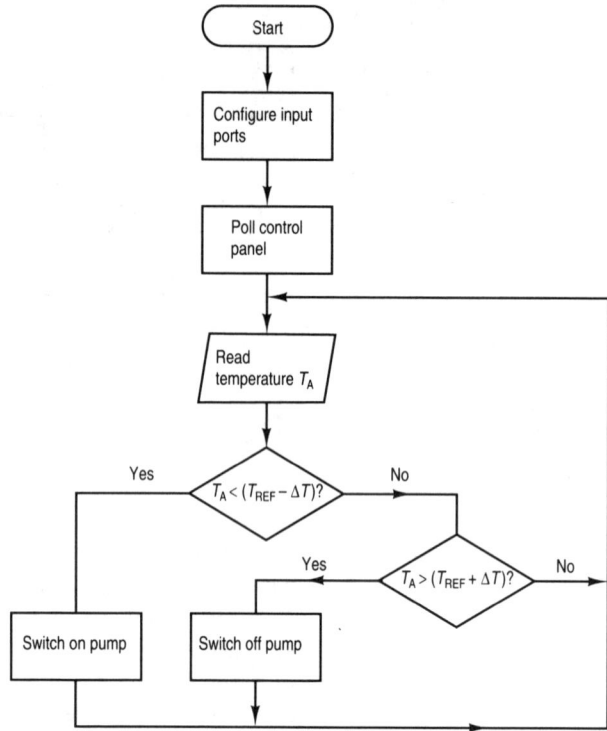

Fig. 7.25 Flow chart of a central-heating system.

8 Assembly language programming

Other than at the engineering level, assembly language programs are not particularly useful since a high-level language compiler is much easier to use and in most cases produces perfectly adequate results. The relative merits of high-level and low-level languages have been discussed in Chapter 3. For many engineering applications, such as microcontrollers in embedded systems, the use of the ANSI C language is widespread since a program is then portable, it is relatively easy to read and understand, it is easier to debug, and it encourages the use of modular programming. Small programs, subroutines, and programs where the utmost speed of execution and/or efficiency are of prime importance, are still commonly written in assembly language. The basic concepts of programming using an assembly language were introduced in Chapter 3 using a simplified basic instruction set. In this chapter an introduction is given to some of the more important instructions used in some commonly employed microprocessors. An 8-bit microprocessor, the Motorola 6809, and some 16-bit microprocessors, the Intel 8086/8088/80286 and the Motorola 68000, are considered. The 8-bit device has been very widely used in the past and it is still used in many equipments today. The Intel 8086, the 8088 and the 80286 have all been employed in older PCs but newer PC designs have used the 32-bit processors 80386 and 80486. These processors all use, more or less, the same instruction set; certainly any program written to run using, say, an 8086 will also run using an 80386 or a 80486. Similarly, the Motorola 68000 has been developed into 32-bit microprocessors such as the 68020 and the 68030 and these all use the same instruction set. The 68000 is widely employed in embedded microelectronic systems while the 68020/68030 are both used in Apple Macintosh computers and in various industrial control applications.

Instruction sets

The instructions that a microprocessor is able to carry out under the control of a program consist of an **op-code** followed by an **operand**. The op-code defines the type of operation that is to be carried out and the operand indicates the address of the memory location which holds the data to be used. In an 8-bit microprocessor the operand may be held in one, or two, memory locations immediately following the location that contains the op-code. 16-bit and 32-bit microprocessors only employ one memory location to hold the operand. A few instructions, such as HALT, consist of the op-code only since then the address is **implied**.

The list of instructions that a microprocessor is able to understand and execute is known as its **instruction set**. Instruction sets vary from one microprocessor to another but they all have certain features in common; they all have data transfer, arithmetic and logic, and branching instructions.

Data transfer instructions

A **data transfer instruction** involves the movement of data between a register and a memory location, or between two registers. In the basic instruction set given on p. 000 the two data transfer instructions are LOAD and STORE. Equivalent instructions to these are provided in all practical microprocessor instruction sets.

Motorola 6809

The 6809 microprocessor has two 8-bit accumulators, A and B, which can be combined to give a single 16-bit accumulator D.

LOAD

The mnemonic LDA is used to indicate 'load accumulator A' while LDB means 'load accumulator B'. Several different addressing modes are available but only the three simplest of them will be considered.

1. **Immediate addressing**. Immediate addressing means that an accumulator is loaded with a constant number and this is indicated to the microprocessor by preceding the number with the symbol #. Thus LDA #14 means 'load accumulator A with decimal number 14'. If a hexadecimal number is to be loaded then this is indicated to the microprocessor by the symbol $. Thus LDB #$14 means 'load accumulator B with hex number 14'. The idea is illustrated in Fig. 8.1 in which the LDA op-code is held in memory location 0300 and the decimal number 14 is held in location 0301. Before the instruction is executed the accumulator is assumed to hold decimal number 06. When the instruction has been executed number 14 has been taken from the memory location and placed into the accumulator, the number 06 previously held in the accumulator is lost but the memory location still holds 14. In this case the memory location is the **source** and the accumulator is the **destination**.

2. **Extended addressing**. With extended addressing the operand consists of the address of the memory location in which the data is held. Thus LDA $202F means 'load the A accumulator with

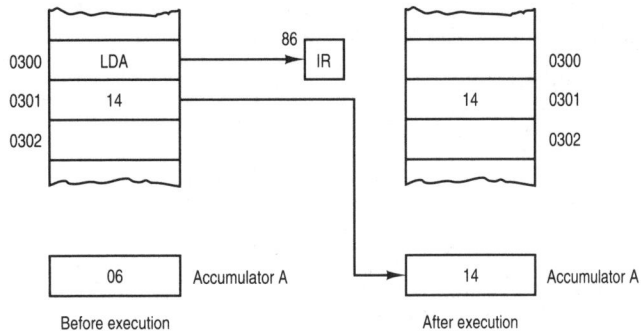

Fig. 8.1 Action of a microprocessor to execute the instruction LDA #14.

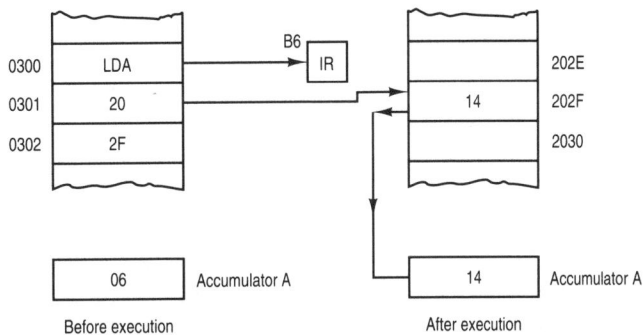

Fig. 8.2 Action of a microprocessor to execute the instruction LDA $202F.

the contents of the memory location whose address is hex 202F'. This particular example of extended addressing is shown in Fig. 8.2 in which memory location hex 202F is supposed to hold decimal number 14. It can be seen that two memory locations are required to hold the address in the case of an 8-bit microprocessor but only one location is required for a 16-bit (or a 32-bit) device.

3. **Direct addressing**. If the memory location is in the **zero page** of the microprocessor's memory map then a form of absolute addressing known as **direct** addressing may be used. This means that the high byte of the address, i.e. 00, need not be given. An example of direct addressing is LDA $80; this means 'load the A accumulator with the contents of memory location $0080'. Direct addressing is shown in Fig. 8.3 in which the address of the memory location used is $0080 and the data stored at this location is decimal number 14.

Similar instructions exist for loading data into the X and Y registers and these instructions may use any of the three addressing modes just mentioned. For example: LDX #$22 means 'load the X register with hex number 22' and LDY $202F means 'load the Y register with the contents of memory location hex 202F'.

STORE
To transfer the data held in the accumulator to either a memory

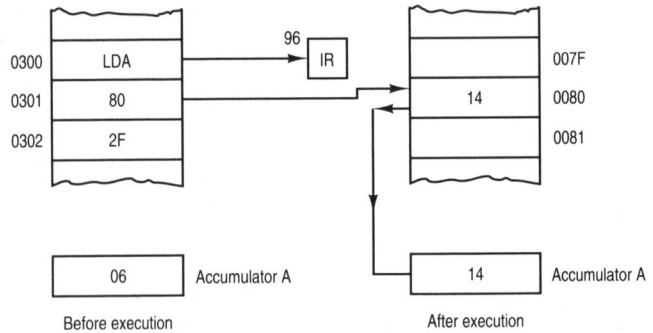

Fig. 8.3 Action of a microprocessor to execute the instruction LDA $80.

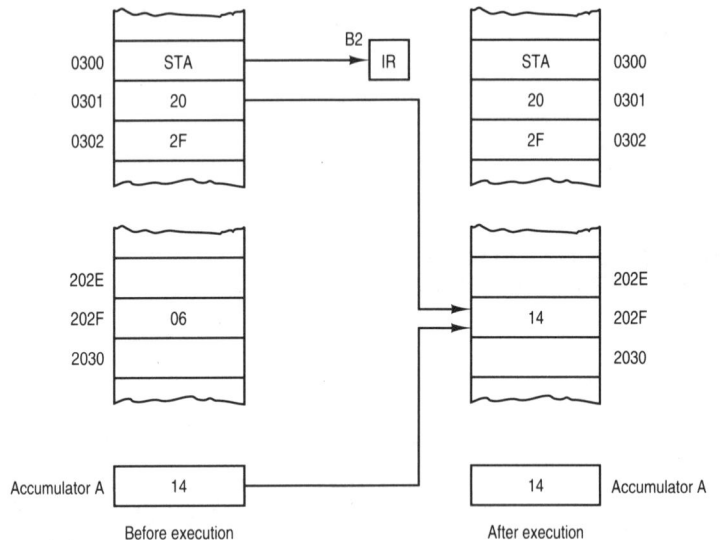

```
0300  LDA      96  IR                           007F
0301  80                        14              0080
0302  2F                                        0081

      06   Accumulator A              14   Accumulator A
   Before execution              After execution
```

Fig. 8.4 Action of a microprocessor to execute the instruction STA $202F.

```
0300  STA      B2  IR          STA   0300
0301  20                       20    0301
0302  2F                       2F    0302

202E                                 202E
202F  06                       14    202F
2030                                 2030

Accumulator A   14         14   Accumulator A
   Before execution        After execution
```

location or to a register the store instruction is used. The immediate addressing mode is not now available.

1. **Extended addressing**. STA $202F means 'store the contents of accumulator A in the memory location whose address is hex 202F' and STX $202F means 'store the contents of the X register in memory location 202F'. The STA operation is shown in Fig. 8.4. Before the instruction is executed the memory location 202F holds 06H; after execution the contents of the accumulator are unchanged but the data held in memory location 202F has been changed to 14H.

2. **Direct addressing**. As before, this form of addressing refers to a location in page zero of the memory map of the processor. The instruction STB $2F, for example, means 'store the contents of accumulator B at the memory location $002F'.

The machine codes for each of these instructions are given in Table 8.1, and clearly they are much less meaningful than mnemonics.

Table 8.1

	Immediate addressing	Extended addressing	Direct addressing
LDA	86	B6	96
LDB	C6	F6	D6
LDX	8E	BE	9E
LDY	10	10	10
STA	—	B2	97
STB	—	F7	D7
STX	—	BF	9F
STY	—	10	10

A knowledge of the machine code versions of each mnemonic is necessary if hand encoding of a program is to be carried out, but this is a tedious and error-prone process. In most cases an assembler is available and this program will carry out all the necessary steps in the assembly of the source code. It is, however, necessary, as discussed in Chapter 4, to allocate memory locations for the start of the program and for each symbolic label that is used.

The input and output ports of this microprocessor are **memory mapped**, which means that they are treated exactly the same as memory locations and data is taken from a port, or is passed to a port, using the LDA (or LDB) and STA (or STB) instructions.

Intel 8086/8088/80286/80386/80486

The 8086/8088/80286 microprocessors are 16-bit devices that have compatible instruction sets. The 80386, 80486 and the new Pentium 32-bit microprocessors are also software compatible with the same instruction set. Compatibility means that they are able to run any programs that have been written for an 8086/8088/80286 processors although the converse is not always true. The 80386/80486 microprocessors contain eight 16-bit registers, four of which are data registers that can also act as accumulators while the other four are index registers. The data registers are labelled as AX, BX, CX, and DX. Register AX is normally used as the accumulator. These Intel processors do not use the instructions LOAD and STORE; instead a general instruction MOVE is used. The general format of the MOVE instruction is MOV destination, source. Hex numbers are indicated to the processor by the letter H at the end of a number and the $ sign is not used. The two basic addressing modes are **immediate** and **direct**.

1. **Immediate addressing**. In the immediate addressing mode the format used is MOV destination, source; thus MOV AX, 15 means 'load register AX with the decimal number 15', and MOV BX, 15H means 'load register BX with hex number 15'. These two instructions are the equivalent of LOAD AX or LOAD BX.

2. **Direct addressing**. In the direct addressing mode the address of a memory location is given in hex form. The instruction MOV AX, 202FH means 'load the AX register with the contents of memory location hex 202F'.

The MOVE instruction is also used to store the data held in a register in a memory location and again the 'destination, source' format is employed. The instruction MOV 202FH, AX means 'store the contents of the AX register in memory location hex 202F'.

If data is to be inputted from an input port then the IN instruction must be employed. The data will only be an 8-bit word so that the destination will be either the high-byte AH, or the low-byte AL, part of the AX register. The instruction will be of the form IN AL, Port, or IN AH, Port, where Port is the address of the input port used. If data is to be outputted from a port then a similar instruction should be used except that OUT is used in place of IN.

Motorola 68000

The Motorola 68000 microprocessor uses a 24-bit address bus, a 16-bit data bus, and 32-bit registers. It is described as either a 16-bit or a 32-bit device. It contains eight data registers, D0 to D7, and seven address registers, A0 to A6. The movement of data is achieved using a MOVE instruction. The format employed is MOVE source, destination [note that this is the other way around to the format used with the 8086, etc., family of microprocessors]. The MOVE mnemonic must be followed by .B (to indicate an 8-bit byte), or by .W (to indicate a 16-bit word), or by .L (to indicate a 32-bit long word). Some examples of the use of this instruction are: MOVE.W $5400, D2 means 'load register D2 with the contents of memory location hex 5400'; MOVE.B #6, D4 means 'load decimal number 6 into register D4'; MOVE.W A3, D3 means 'load the register D3 with the contents of register A3'. There are no IN or OUT instructions in the instruction set so the input/output ports must be memory mapped.

Arithmetic instructions

The four arithmetic instructions are ADD, SUBTRACT, MULTIPLY, and DIVIDE. All 8-bit microprocessors are able to add or subtract two 8-bit numbers, i.e. numbers up to 255. The addition of larger numbers can also be carried out but the method that is used is beyond the scope of this book. 16-bit and 32-bit microprocessors can, of course, operate on much larger numbers. One of the numbers to be added or subtracted is placed in an accumulator and the other number is taken from another register or from memory. The ADD or the SUBTRACT instruction is then executed. The result of the computation is stored in the accumulator. If the result of a subtraction is negative the answer will appear in what is known as 'two's complement' form. Many 8-bit microprocessors are not provided with MULTIPLY or DIVIDE instructions but all 16-bit and 32-bit microprocessors do include these instructions in their instruction sets.

Motorola 6809

The ADD instruction is obtainable either with or without carry and the three addressing modes — immediate, extended, and direct — are all available. The ADCA and ADCB instructions take account of any previous carry while ADDA and ADDB do not. Figures 8.5(a) and (b) show the contents of accumulator A and a memory location before and after an ADDA instruction has been carried out using direct addressing. Initially the accumulator holds hex number 06 and the memory location $0084 holds hex number 04; after execution of the ADD instruction the accumulator holds number 0AH and the contents of the memory location are unchanged.

The SUBTRACT instruction is also obtainable with or without carry, i.e. as SUB or as SBC, using the same addressing modes. Figure 8.6 shows the situation when the SUBTRACT instruction is carried out, and once again the contents of the memory location are left unaltered.

The COMPARE instruction is basically a SUBTRACT instruction in which the difference is not stored in an accumulator and so the contents of the accumulator are not altered. The result of the subtraction does, however, affect a flag which is used in conjunction with a branch instruction. The COMPARE instruction is provided in a number of different ways using direct, extended, or immediate addressing. Two of these ways are:

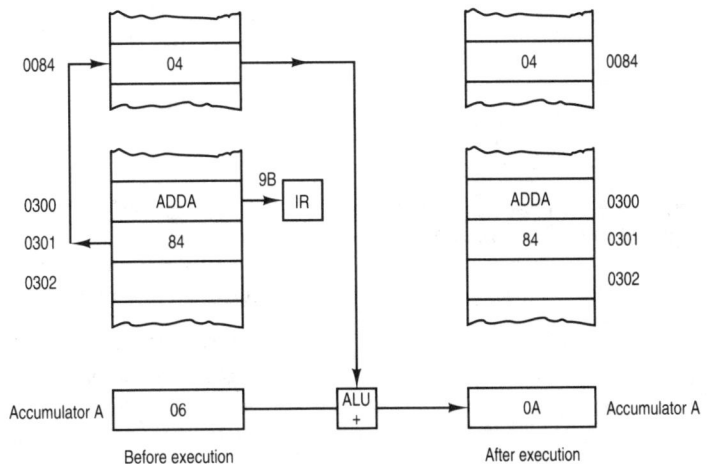

Fig. 8.5 Action of a microprocessor to execute the instruction ADDA, $0084.

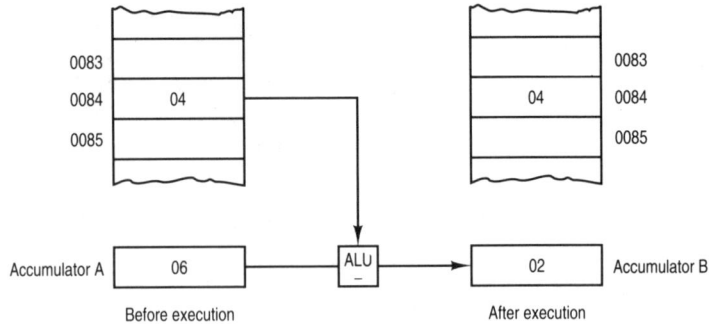

Fig. 8.6 Action of a microprocessor to execute the instruction SUBA, $0084.

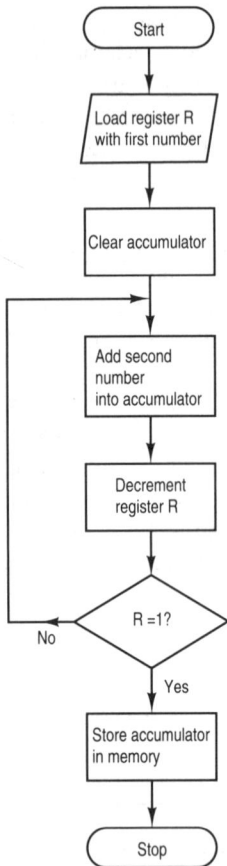

Fig. 8.7 Flow chart to multiply together two numbers A and B.

- CMPA (or CMPB): compare the contents of a memory location with the contents of the accumulator
- CMPX (or CMPY): compare the contents of a memory location with the contents of the X (or Y) register.

Instructions also exist that cause the contents of an accumulator or of a memory location to be either incremented or decremented. Incremented means to ADD 1, and decremented means to SUBTRACT 1, from the contents of an accumulator or a memory location. Note that if a register holds all 1s and is incremented it will then hold all 0s, and if the register when holding all 0s is decremented the result is all 1s.

INCA means 'increment the A accumulator'; DECA means 'decrement the A accumulator'; while INC/DEC mean increment/decrement the contents of a memory location. When the contents of a memory location are to be incremented or decremented the contents of that location are moved into an accumulator. The contents are incremented or decremented and the amended data is then moved back into the same memory location.

If an 8-bit microprocessor does not include either a multiply or a divide instruction then a subroutine must be employed for these purposes. Multiplication can be carried out by repeated addition and Fig. 8.7 shows the flow chart for such an operation where the number A is multiplied by another number B. Figure 8.8 shows the flow chart for a subroutine that will divide a number A by another number B, store the quotient in memory location C and the remainder in location D.

Intel 8086/8088/80286/80386/80486

These 16-bit and 32-bit microprocessors include the following arithmetic instructions:

ADD destination, source: add two numbers
ADC destination, source: add with carry two numbers
SUB destination, source: subtract one number from another
SBB destination, source: subtract with borrow

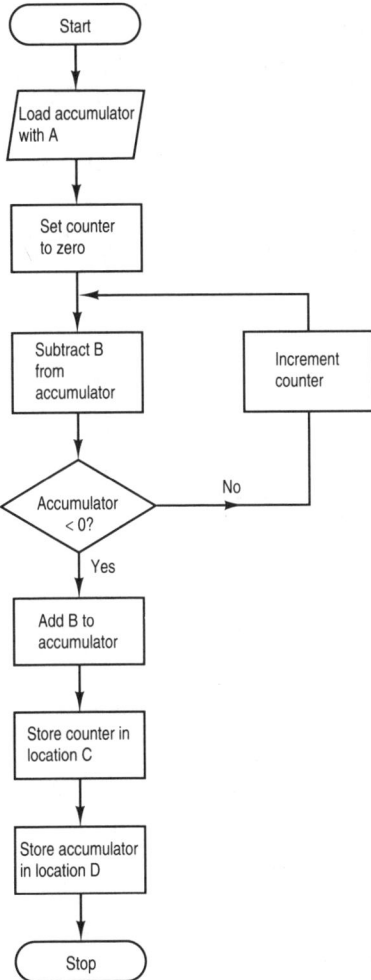

Fig. 8.8 Flow chart to divide A by B.

INC destination: increment either a register or memory

DEC destination: decrement either a register or memory

MUL source: multiply

DIV source: divide.

Motorola 68000

The arithmetic instructions in the 68000's instruction set include:

ADD source, destination: add binary
ABCD source, destination: add BCD
SUB source, destination: subtract binary
SBCD source, destination: subtract BCD
CMP source, data register: compare two numbers
DIVS source, data register: divide signed numbers
DIVU source, data register: divide unsigned numbers
MULS source, data register: multiply signed numbers
MULU source, data register: multiply unsigned numbers.

Test and branch instructions

A **conditional branch** or **jump** instruction makes the execution of a program move away from its normal step-by-step sequence and, instead, go to another specified point in the program *if* some pre-stated condition has been satisfied. An **unconditional branch** or **jump** will always make the program move to some other specified point.

Conditional branch

A few of the various branching instructions provided in the various instructions sets are shown by Table 8.2.

These branch instructions are illustrated in Fig. 8.9. If the memory location that is branched to is located earlier in the program than the point where the branch occurs a loop is formed and two examples of this are shown in Fig. 8.10.

Table 8.2

	BASIC equivalent	6809	8086, etc.	68000
Branch if minus	IF N < 0 GOTO	BMI	JM	BMI
Branch if not 0	IF N ≠ 0 GOTO	BNE	JNZ	BNE
Branch if plus	IF N ≥ 0 GOTO	BPL	JP	BPL
Branch if 0	IF N = 0 GOTO	BEQ	JZ	BEQ

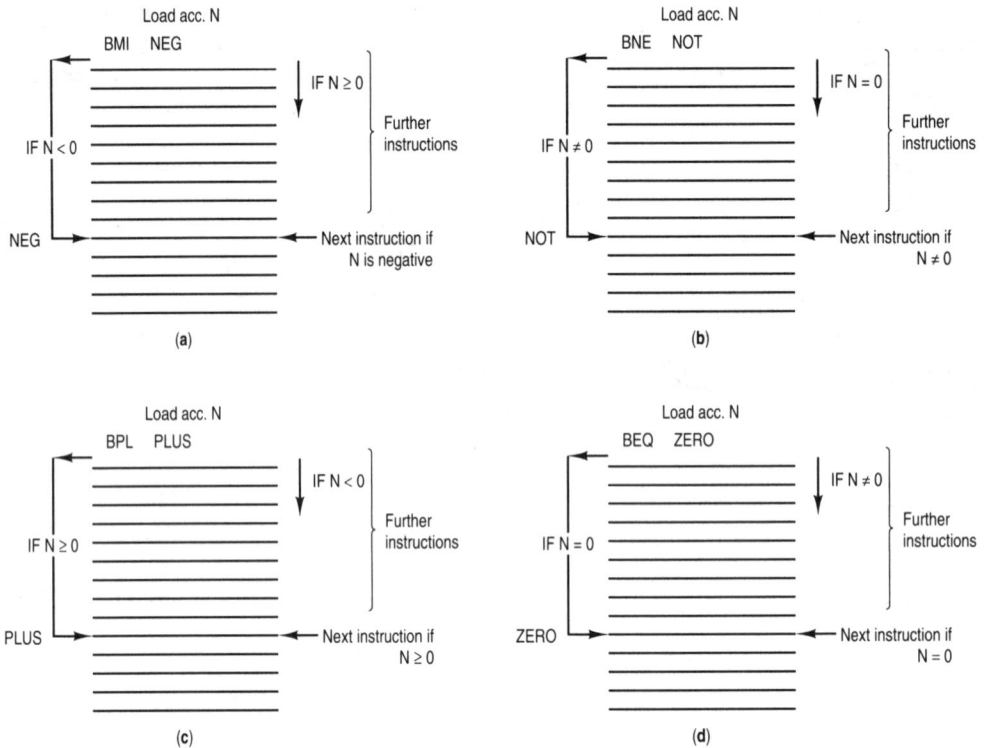

Fig. 8.9 Branch instructions: (a) BMI, (b) BNE, (c) BPL, and (d) BEQ.

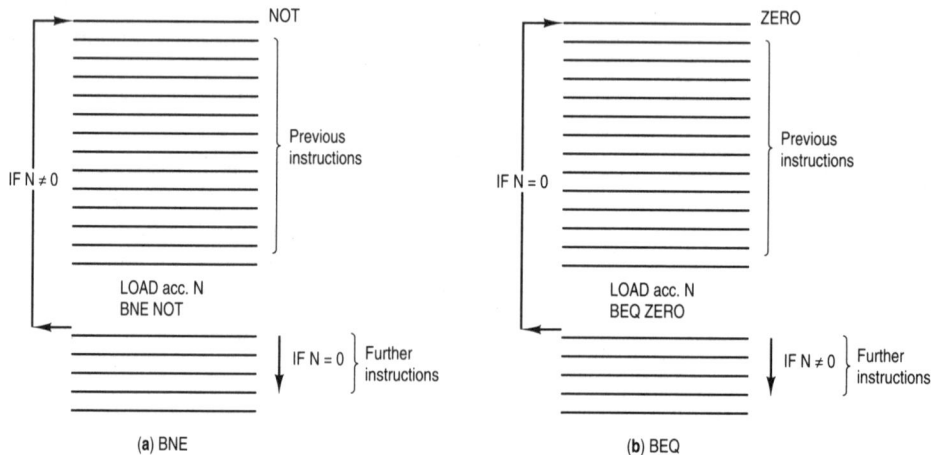

Fig. 8.10 Branch instructions forming a loop: (a) BNE and (b) BEQ.

It is usually better to not use the BPL or BMI instructions after a COMPARE instruction because an error may possibly occur with certain numbers. Instead, in this situation it is safer to use two other instructions BCS and BCC. BCS is 'branch if carry set' and this instruction gives a branch if a result is greater than or equal to zero,

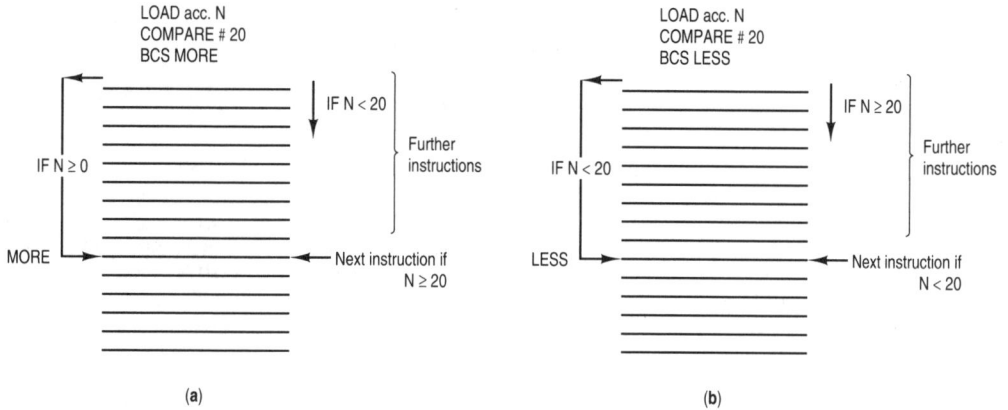

Fig. 8.11 Branch instructions: (a) BCS and (b) BCC.

and BCC is 'branch if carry clear' and now a branch occurs if the result is less than zero. These two instructions are illustrated in Fig. 8.11.

Unconditional branch

Unconditional branch/jump instructions include JMP for all devices, JSR (jump to subroutine) for all except the Intel devices, and CALL (jump to subroutine) for the Intel devices.

The JSR or CALL instructions are used to move the execution of a program to a subroutine. When an instruction that calls a subroutine is executed the contents of the program counter are saved and the operand of the subroutine is placed into the program counter. The program then jumps to the first instruction of the subroutine. At the end of the subroutine another instruction, generally either RTS (return from subroutine) or RET (return), causes the saved program counter value to be put back into the program counter so that the next instruction to be executed is the one that immediately follows the subroutine call. The concept of calling a subroutine was shown in Fig. 4.27.

Assembly language programs

To illustrate the use of an instruction set a number of simple assembly language programs follow. The instructions used are limited to those mentioned earlier, which means that there is often another, probably more efficient, way in which a program could be written.

Example 8.1

Write a program to determine the sum of the two numbers 5 and 7 and to store the sum in memory location $0080. Start the program in location $0200.

Solution
SUM EQU $0080
ORG EQU $0200

	6809	8086, etc.	68000/20/30
LOAD acc. #5	LDA #5	MOV AX, 5	MOVE.B #5, D0
ADD acc. #7	ADD #7	ADD AX, 7	ADD #7, D0
STORE acc. SUM	STA SUM	MOV SUM, AX	MOVE.B AX, SUM
STOP	END	HLT	END

Example 8.2

Write a program to calculate the sum of the two numbers held in memory locations $0080 and $0081 and to store the result in location $0082.

Solution
ORG $0200

	6809	8086, etc.	68000/20/30
LOAD acc. $0080	LDA $80	MOV AX, 80H	MOVE.B $80, D0
ADD $0081	ADDA $81	ADD AX, 81H	ADD $81, D0
STORE acc. $0082	STA $82	MOV $82, AX	MOVE.B D0, $82
STOP	END	HLT	END

In an 8-bit microprocessor the result of an ADD instruction must not be larger than 255 or an error will occur. For a 16-bit microprocessor this maximum number is 65535. The potential error can be avoided but the means of so doing are beyond the scope of this book.

Example 8.3

Write an assembly-language program to compare a byte of data held in memory location $0300 with another byte of data that is held in location $0301. If the two values are equal, this should be indicated by placing $00 in location $0302; if the values are not equal, location $0302 should be loaded with $FF.

Solution
ORG EQU $0200

		6809		8086, etc.		68000/20/30
	LOAD acc. $0300	LDA $0300		MOV AX, 0300H		MOVE.B $0300, D0
	COMPARE $0301	CMPA $0301		CMP 0301H, AX		SUB $0301, D0
	BEQ ZERO	BEQ ZERO		JZ ZERO		BEQ ZERO
	LOAD acc, #FF	LDA #$FF		MOV AX, FFH		MOVE.B $FF, D0
	STORE acc, $0302	STA $0302		MOV 0302H, AX		MOVE.B D0, $0302
ZERO	LOAD acc. #$00	ZERO LDA #$00	ZERO	MOV AX, 00H	ZERO	MOVE.B $00, D0
	STORE acc. $0302	STA $0302		MOV 0302H, AX		MOVE.B D0, $0302
	STOP	END		HLT		END

If the MULTIPLY instruction is available then a subroutine that uses repeated addition, as shown by the flow chart of Fig. 8.7, is not required. An example of the use of the MULTIPLY instruction is given in Example 8.4.

Example 8.4

Write a program to perform the following calculation:

$$x = a^2 + b^2 + 4ac$$

If the result is positive store 1 in location $0250 and if the result is negative store 0 in that location. Start the program in location $0200.

Solution

```
a    EQU $0260
b    EQU $0261
c    EQU $0262
x    EQU $0263
a²   EQU $0264
b²   EQU $0266
PN   EQU $0250
ORG  $0200
```

	6809		8086, etc.		68000/20/30
	LOAD acc. a		LDA a		MOV AX, a		MOVE.W a, D0
	MULTIPLY a		MUL a		MUL a		MULU.W a
	STORE acc. a^2		STA a^2		MOV AX, a^2		MOVE.W a^2, D0
	LOAD acc. b		LDA b		MOV AX, b		MOVE.W b, D0
	MULTIPLY b		MUL b		MUL b		MULU.W b
	STORE acc. b^2		STA b^2		MOV AX, b^2		MOVE.W b^2
	LOAD acc. a		LDA a		MOV AX, a		MOVE.W a, D0
	MULTIPLY 4		MUL #4		MUL #4		MULU.W 4
	MULTIPLY c		MUL c		MUL c		MULU.W c
	ADD a^2		ADDA a^2		ADD AX, a^2		ADD.W a^2, D0
	ADD b^2		ADDA b^2		ADD AX, b^2		ADD.W a^2, D0
	STORE acc. x		STA x		MOV AX, x		MOVE.W x, D0
	BPL POS		BPL POS		BPL POS		BPL POS
	LOAD acc. #0		LDA #0		MOV AX, #0		MOVE.W #0, D0
	JUMP NEG		JMP NEG		JMP NEG		JMP NEG
POS	LOAD acc. #1	POS	LDA #1	POS	MOV AX, #1	POS	MOVE.W #0, D0
NEG	STORE acc. PN		STA PN		MOV PN, AX		MOVE.B D0, PN
	STOP		END		HLT		END

Example 8.5

Draw the flow chart and then write a program to check a data word held in memory location $0250. If the word is negative $FF should be stored in location $0251; if it is zero then $00 should be stored in $0251; and should the word be positive then $01 must be placed into location $0251.

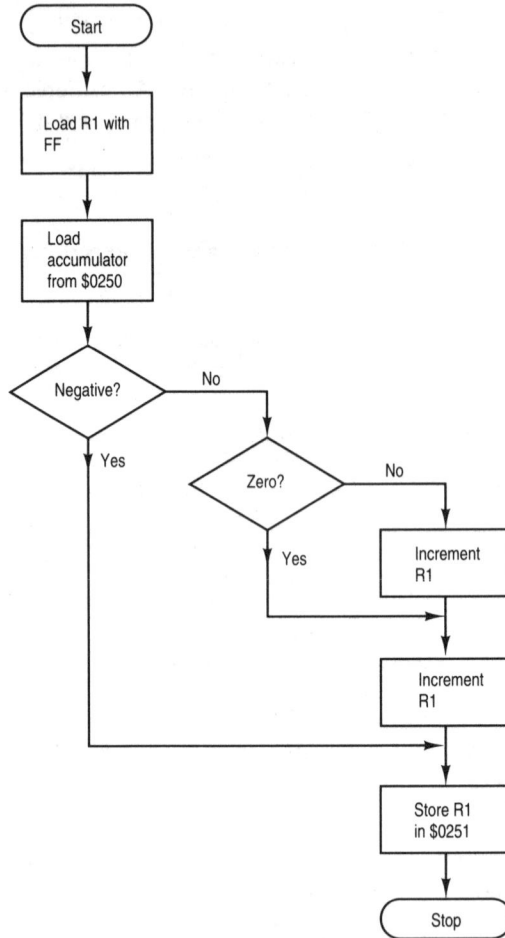

Fig. 8.12 Example 8.5: flow chart.

Solution

The flow chart is shown in Fig. 8.12. The program is:

ORG EQU $0200

		6809		8086, etc.		68000/20/30
	LOAD reg. #$FF	LDB #$FF		MOV AX, #FFH		MOVE.B #$FF, D0
	LOAD acc. $0250	LDA $0250		MOV BX, $0250H		MOVE.B $0250, D1
	BMI NEG	BMI NEG		JM NEG		BMI NEG
	BEQ ZERO	BEQ ZERO		JZ ZERO		BEQ ZERO
	INCR. X	INCB		INC AX		ADD #1, D0
ZERO	INCR. reg.	ZERO INCB	ZERO	INC AX	ZERO	ADD #1, D0
NEG	STORE X $0251	NEG STB $0251	NEG	MOV 0251H, AX	NEG	MOVE.B D0, $0251
	STOP	END		HLT		END

Example 8.6

Two numbers are stored in memory locations $03DD and $03DE. If the smaller number is held in $03DE interchange the contents of the two locations.

Solution
ORG EQU $0020

		6809		8086, etc.		68000/20/30
	LOAD acc. $03DE	LDA $03DE		MOV AX, 03DEH		MOVE.B $03DE, D0
	COMPARE $03DD	CMPA $03DD		CMP 03DDH, AX		SUB $03DD, D0
	BMI SWOP	BMI SWOP		JM SWOP		BMI SWOP
	STOP	END		HLT		END
SWOP	TRANSFER acc/R	SWOP LDB $03DD	SWOP	MOV BX, 03DDH	SWOP	MOVE.B $03DD, D1
	LOAD acc. $03DD	STA $03DD		MOV 03DDH, AX		MOVE.B D0, $03DD
	STORE acc. $03DE	STB 03DE		MOV 03DEH, AX		MOVE.B D1, $03DE
	STOP	END		HLT		END

Example 8.7

Write a program to calculate the sum and the difference of the two numbers that are stored in memory locations $0200 and $0201. Their sum should be stored at location $0250 and their difference at $0251.

Solution
ORG EQU $0200

	6809	8066, etc.	68000/20/30
LOAD acc. $0200	LDA $0200	MOV AX, 0200H	MOVE.B $0200, D0
ADD $0201	ADD $0201	ADD AX, 0201H	ADD $0201, D0
STORE acc. $0250	STA $0250	MOV 0250H, AX	MOVE.B D0,$0250
LOAD acc. $0200	LDB $0200	MOV BX, 0200H	MOVE.B $0200, D1
SUBTRACT $0201	SUBB $0201	SUB BX, 0201H	SUB $0201, D1
STORE acc. $0251	STB $0251	MOV 0251H, BX	MOVE.B D1, $0251
STOP	END	HLT	END

Example 8.8

Write a program to implement the multiplication subroutine whose flow chart is given in Fig. 8.7.

Solution
A   EQU $0250
B   EQU $0251
RES   EQU $0252
ORG   EQU $0200

		6809		8086, etc.		68000/20/30
	LOAD R1, A	LDB, A		MOV AX, A		MOVE.B A, D0
	LOAD acc. #0	LDA #$00		MOV BX, $00H		MOVE.B #$00
AGAIN	ADD acc., B	AGAIN ADDA, B	AGAIN	ADD BX, B	AGAIN	ADD.B, D1
	DECREMENT R1	DECB		DEC AX		SUB#1, D0
	BNE AGAIN	BNE AGAIN		BNE AGAIN		BNE AGAIN
	STORE RES	STA RES		MOV RES, BX		MOVE.B D1, RES
	STOP	HLT		END		HLT

Example 8.9

Draw the flow chart and write a program to produce all the positive integer powers of 2, i.e. 2^n.

Solution
The flow chart is shown in Fig. 8.13 and the program is:
RES EQU $0250
ORG $0200

	6809		*8086, etc.*		*68000/20/30*
	LOAD R1 #n		LDB #n		MOV AX, n		MOVE.W D0, #n
	LOAD acc. #2		LDA #2		MOV BX, #2		MOVE.W D1, #2
LOOP	MULTIPLY #2	LOOP	MUL 2	LOOP	MUL BX, #2	LOOP	MULU.W D1, #2
	DECREMENT R1		DECB		DEC AX		SUBI #1, D0
	COMPARE #1		CMP #1		CMP #1		CMP #1
	BNE LOOP		BNE LOOP		BNE LOOP		BNE LOOP
	STORE acc. RES		STA A RES		MOV RES, BX		MOVE.W D1, RES
	STOP		HLT		END		HLT

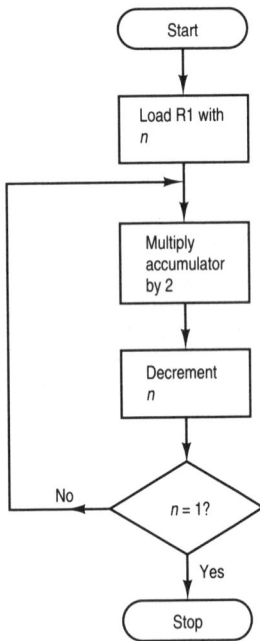

Fig. 8.13 Example 8.9: flow chart.

Example 8.10

Write a program to implement the flow chart given in Fig. 4.14(a).

Solution
The figure shows the flow chart of an AND gate. The address of the input port must be known and is $2000. Any lines of the 8-line port that are not used for an input should be set to binary 1. The output port may have all its lines set as outputs and its address is $3000.

The first instruction must read in the data at the input port and load it into an accumulator. The first instruction is LOAD $2000. It is then necessary to test to see if all the inputs are at the binary 1 voltage level. If they are then adding 1 will make all lines go to 0 [(11111111) +1 = 100000000]. If all bits in the accumulator are 0 after an arithmetic operation a flag is set. Thus the second instruction must be INCREMENT accumulator. If the zero flag is set the program should jump to a new point labelled, say, ONES. Hence the third instruction is BEQ ONES. If the zero flag is not set then the output port must be set to binary 0. First load the accumulator with $00, i.e. LOAD accumulator £$00. Then store the contents of the accumulator in memory location $3000 and return to the beginning of the program. Thus the next instructions are STORE accumulator $3000 and JUMP START. If the zero flag was set then the output port must be set to binary 1, so load the accumulator with FF and then store this value in location $3000. The next two instructions are therefore LOAD accumulator #$FF, and STORE accumulator $3000. Lastly, this instruction should be followed by a return to the beginning of the program and so the final instruction is JUMP START. The program is:

ORG EQU $0200

	6809		*8086, etc.*		*68000/20/30*
START	LOAD $2000	START	LDA $200	START	MOV AX, 2000H	START	MOVE.B $2000, AX
	INCREMENT acc.		INCA		INC AX		ADD #$01
	BEQ ONES		BEQ ONES		JZ ONES		BEQ ONES

```
         LOAD acc. #$00          LDA #$00          MOV AX, 00          MOVE.B #$00, AX
         STORE acc. $3000        STA $3000         MOV 3000H, AX       MOVE.B AX, $3000
         JUMP START              JMP START         JMP START           JMP START
ONES     LOAD acc. #$FF    ONES  LDA #$FF    ONES  MOV AX, FFH   ONES  MOVE.B $FF, AX
         STORE acc. $3000        STA $3000         MOV 3000H, AX       MOVE.B AX, $3000
         JUMP START              JMP START         JMP START           JMP START
```

Embedded systems, particularly those that employ a micro-controller, have a very specific set of requirements. One version of the C language has been standardized by the American National Standards Institute (ANSI) and is often known as ANSI C. The use of ANSI C ensures that a program written in that language is both portable from one C compiler to another, is easily able to access all input/output ports, and has only about a 20% reduction in speed compared with an assembly language. C is written in English-like statements that make the operation of the code fairly easy to understand, even without any comments. C also encourages modular programming and this avoids the need to write a subroutine each time a new program is written.

9 Information transmission

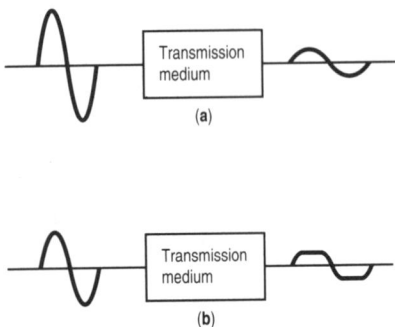

Fig. 9.1 Effects of (a) attenuation and (b) distortion on a sinusoidal waveform.

There is often a need for information generated, or stored, at one location to be transmitted to some other location. The two sites in question may be situated near each other or they may be some considerable distance apart. The information, which may be in analogue or digital form, must be applied to a transmission medium that is able to transmit the signal. The media employed are generally either some kind of cable or a radio link. The cables that are employed may either have copper conductors or be of the optical fibre type. Telecommunication cables and radio links provide an important part of a **telecommunication system** that may be employed for the transmission of information. The information signals that are transmitted could be any of the following: radio, television, data, telegraphy or telephony. In all cases the information signal will need to be processed before it is transmitted because all types of transmission systems are subject to various limitations. The limitation may be one of restricted bandwidth, or an inability to transmit any d.c. component, or a maximum bit rate that can be transmitted, or added noise and distortion. Whatever the transmission medium that is employed the information signal will suffer loss, or **attenuation**, as it travels from the transmitting end of the circuit to the receiving end. The effect of this attenuation on a signal reduces its amplitude and the effect of any distortion that may be introduced is to alter the signal waveform. The effects of attenuation and distortion on a signal are shown in Fig. 9.1. In figure (a) a sinusoidal input signal has its amplitude reduced by its transmission over the medium and in figure (b) distortion has changed the shape of the signal so that it is no longer sinusoidal. In all cases it is necessary that sufficient information is available at the receiving end of a communication link for the intelligence to be used and, in the case of music and television, to be enjoyed. The demands made upon the transmission medium depend upon the type of signal being transmitted; for speech and music transmissions the requirement is that sufficient frequencies must exist in the received signal for the information to be understood and, perhaps, enjoyed. For the transmission of television pictures it is necessary that the waveforms of the transmitted signals are preserved, otherwise the received picture will be degraded. The waveform of a data signal must also be preserved otherwise it is likely that errors will occur in the received data. It is therefore necessary for the range of frequencies transmitted by various kinds of signals to be known.

Time, distance, and frequency scales

The analogue signal waveforms given in Chapter 6 have all been drawn showing current or voltage plotted to a base of time. This is the most common way of representing a signal waveform but is not the only method possible. Instead, a signal may be represented by plotting voltage (or current) against either distance or frequency.

Time scale

Figure 9.2 shows a sinusoidal voltage wave in which instantaneous voltage has been plotted against time. The wave has a peak value, or amplitude, of V_m volts and a period, or periodic time, of T seconds. The positive and negative peak values, $\pm V_m$, are equal (if they are not equal the wave has been distorted and is no longer a pure sine wave). The frequency f of the wave, i.e. the number of complete cycles that occur in one second, is equal to the reciprocal of the periodic time. Thus,

$$f = 1/T \text{ Hz} \tag{9.1}$$

In practice, most of the signals that are employed in computer/ electronic/telecommunication circuitry and systems have a non-sinusoidal waveform. The examples given in Chapter 6 included rectangular, square, and sawtooth waveforms. The frequency of a non-sinusoidal waveform is equal to the reciprocal of its periodic time — as with a sine wave. This concept works well enough with square, sawtooth, and triangular waves and, for example, the clock waveform used in a computer system is quoted in MHz. The digital signals employed in computer systems use various combinations of 1 and 0

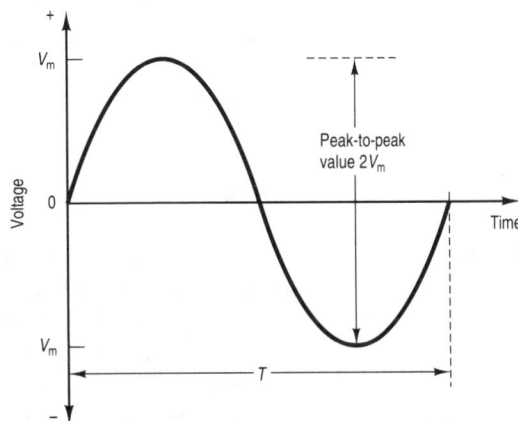

Fig. 9.2 Sinusoidal voltage plotted against a time axis.

(a)

(b)

(c)

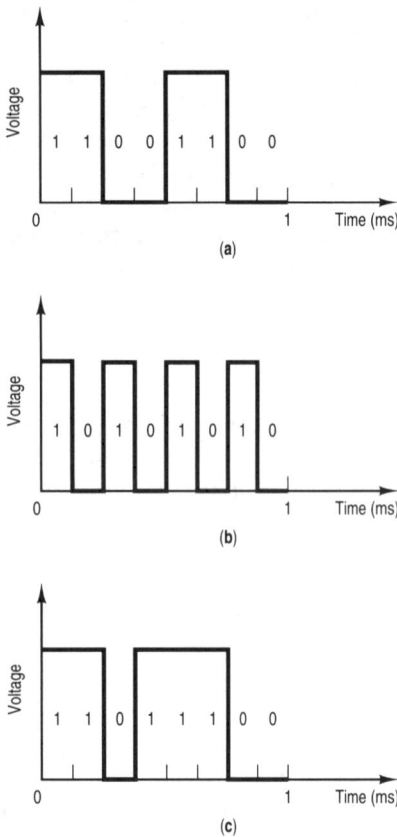

Fig. 9.3 Three digital signals with the same bit rate.

bits, which means that their frequency is continually changing. This fact makes it more convenient, and more usual, to consider instead the **bit rate** of a digital waveform.

The bit rate of a digital waveform is the number of bits that occur in 1 second. Typically, the bit rate may vary from as low as 50 bits/s for very slow data circuits up to several megabits/s for a **local line network**. Figures 9.3(a) and (b) show two digital waveforms that have the same bit rate but signal (b) has twice the frequency of signal (a), while figure (c) shows another digital signal at the same bit rate whose frequency is not constant.

Example 9.1

(a) Calculate the bit rate of each of the three digital signals shown in Fig. 9.3.
(b) Calculate the frequency of each of the first two waveforms.

Solution
(a) Bit duration $= (1 \times 10^{-3})/8 = 1.25 \times 10^{-4}$ s.
      Therefore bit rate $= 1/(1.25 \times 10^{-4}) = 8000$ bits/s. (*Ans.*)
(b) $T = 4 \times 1.25 \times 10^{-4} = 5 \times 10^{-4}$.
      $f = 1/(5 \times 10^{-4}) = 2000$ Hz. (*Ans.*)
      $T = 2 \times 1.25 \times 10^{-4} = 2.5 \times 10^{-4}$.
      $f - 1/(2.5 \times 10^{-4}) = 4000$ Hz. (*Ans.*)

All repetitive non-sinusoidal waveforms consist of the sum of a number of sinusoidal components at a number of different, but related, frequencies (p. 178). The periodic time of the waveform is equal to the periodic time of the lowest frequency sinusoidal component. This component is known as the **fundamental-frequency** component. All the other components in the wave will be at frequencies that are integer multiples of the fundamental frequency. Thus the **second-harmonic** component is at twice the fundamental frequency, the **third-harmonic** is at three times the fundamental frequency, and so on. Usually, the higher the order of a harmonic the smaller is its amplitude. In any non-sinusoidal waveform particular harmonics may, or may not, be present and some of the more important examples of such waveforms are given on p. 125. Also, a d.c. component may exist, and this is equal to the mean (or average) value of the waveform.

Distance scale

When a signal waveform travels through the atmosphere, or over a pair of conductors in a cable, it will have travelled a certain distance in the periodic time of that waveform. This is shown in Fig. 9.4 for a sinusoidal voltage wave. The instantaneous voltage of the wave has a particular value, say the positive peak value, at equal distances that are repeated along the length of the transmission path. This distance

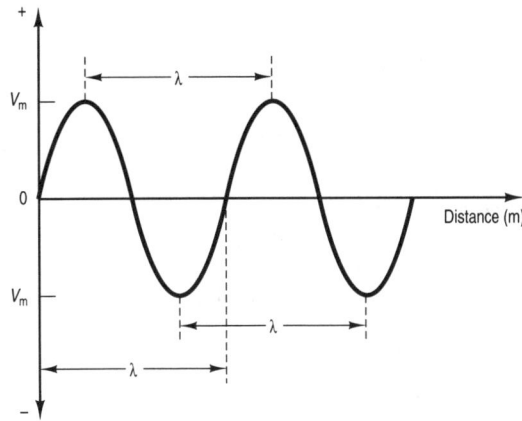

Fig. 9.4 Sinusoidal voltage plotted against a distance axis.

is known as the **wavelength** λ of the signal. The wavelength is the distance travelled by a sinusoidal wave in the time that it takes to go through one complete cycle of instantaneous values.

Velocity is equal to (distance travelled)/(time taken) and for a distance equal to the signal wavelength:

$$\text{velocity} = \text{wavelength}/(\text{periodic time}) = \lambda/T = \lambda f \qquad (9.2)$$

Example 9.2

A 60 MHz radio wave travels through the atmosphere at the velocity of light, i.e. at 3×10^8 m/s. Calculate its wavelength.

Solution
$\lambda = (3 \times 10^8)/(60 \times 10^6) = 5$ m.   (*Ans.*)

Because of the attenuation introduced by the transmission medium, the amplitude of the voltage waveform will become progressively smaller as the wave travels over the medium. The effect of this attenuation upon a sinusoidal waveform is shown in Fig. 9.5.

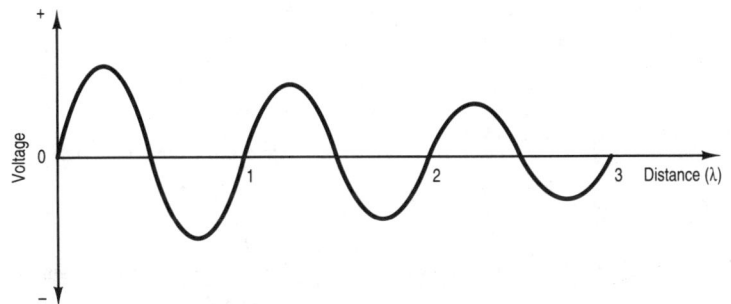

Fig. 9.5 Amplitude of sinusoidal voltage decreases with distance.

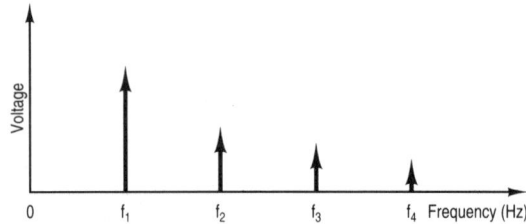

Fig. 9.6 Frequency spectrum of a fundamental plus three harmonics.

Frequency scale

Non-sinusoidal waveforms are sometimes plotted against frequency instead of being plotted against time, which is more usual. This method results in the waveform of the signal not being displayed; instead, all the frequency components of the wave are shown. The display that is obtained is known as the **frequency spectrum** of the waveform. One example is given here (see Fig. 9.6) merely to show the general form of a spectrum diagram; it is for a waveform that consists of a fundamental frequency component plus components at its second, third, and fourth harmonics. Some specific instances of frequency spectrum diagrams are considered later in the chapter.

Synthesis of non-sinusoidal waveforms

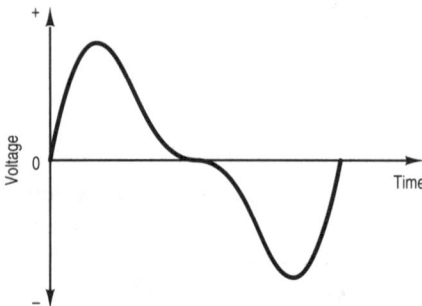

Fig. 9.7 Fundamental plus second harmonic.

If one or more harmonic components are added to a fundamental frequency component, a non-sinusoidal waveform is obtained. The waveform of the non-sinusoidal signal depends upon the order(s) of the harmonic(s) added to the fundamental and their amplitudes relative to the fundamental. Figure 9.7 shows the resultant waveform that is obtained when a second-harmonic component is added to a fundamental; the amplitude of the harmonic is one-half the amplitude of the fundamental. Figure 9.8 shows the effect of adding a third-harmonic component to a fundamental-frequency component, clearly the resultant waveform tends toward square shape. If further odd harmonics are added in turn, firstly the fifth harmonic, then the seventh harmonic, then the ninth harmonic, and so on, the waveshape will increasingly become an even better approximation to a square wave. A perfectly square wave contains a fundamental-frequency component plus all the odd-order harmonics up to a very high number.

Example 9.3

Determine the component frequencies of the square waveform shown in Fig. 9.9.

Solution

The periodic time of the waveform is 1 ms. Therefore its fundamental frequency is

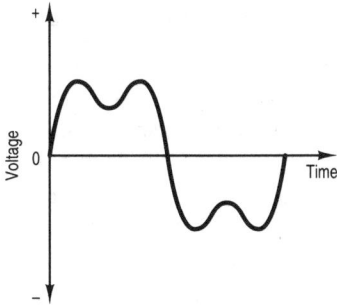

Fig. 9.8 Fundamental plus third harmonic.

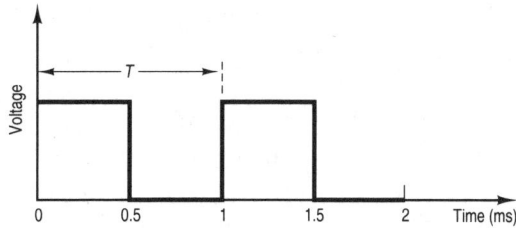

Fig. 9.9 Example 9.3: a square waveform.

$1/(1 \times 10^{-3}) = 1$ kHz. (*Ans.*)

Also present in the waveform are the odd harmonics of 1 kHz, i.e.

3 kHz, 5 kHz, 7 kHz, 9 kHz, etc. (*Ans.*)

A rectangular waveform may contain even- and odd-order harmonic components and these are best determined with the aid of a **spectrum diagram**.

Figure 9.10(a) shows a rectangular waveform that has a periodic time of T and a pulse width of τ. The spectrum diagram of the waveform is a plot of voltage against frequency and it is of the form shown in Fig. 9.10(b). The spectrum diagram consists of a number of lobes inside each of which there are a number of **spectral lines**. The spectral lines are spaced at frequency intervals of $1/T$ and each lobe has a width equal to $1/\tau$. If a spectral line has a frequency equal to that of a lobe zero, then the amplitude of that line is zero. The negative values indicate that at time $t = 0$ the component has a negative voltage.

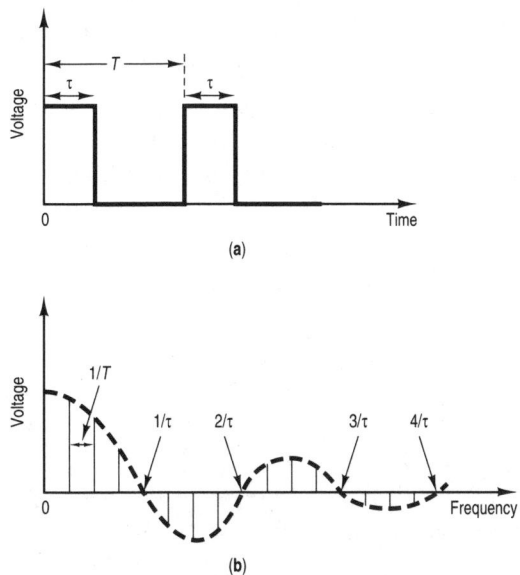

(a)

(b)

Fig. 9.10 (a) Rectangular waveform; (b) spectrum diagram of (a).

The frequency components of the rectangular waveform are:

(a) A d.c. component ($f = 0$) whose amplitude is equal to the average value of the waveform. If the waveform has equal positive and negative voltages then the d.c. component is of zero amplitude.
(b) Components are at frequencies of $1/T$, $2/T$, $3/T$, $4/T$ and so on. Clearly the amplitudes of the components get progressively smaller within each successive frequency lobe.

Example 9.4

A rectangular waveform has a periodic time of 1 ms and each pulse has a duration of 0.25 ms. Determine the frequencies of the components.

Solution
The zeros of the lobes occur at frequencies of $1/\tau = 1/(0.25 \times 10^{-3}) = 4$ kHz and the spectral lines are spaced apart at $1/T = 1/(1 \times 10^{-3}) = 1$ kHz frequency intervals. Hence:

The fundamental frequency is 1 kHz. (*Ans.*)
Also present are components at frequencies of 2 kHz, 3 kHz, 5 kHz, 6 kHz, 7 kHz, 9 kHz, 10 kHz, 11 kHz, and so on. (*Ans.*)

There are no components at 4 kHz, 8 kHz, 12 kHz, etc. because these frequencies correspond to lobe zeros.

In the case of a square waveform the pulse width τ is equal to one-half of the periodic time T and so the spectrum diagram is as shown in Fig. 9.11. It can be seen that there is only one spectral line per lobe and so the square waveform contains components at:

(a) a d.c. component (if the wave is not symmetrical about the zero voltage axis)
(b) a fundamental frequency component $f \stackrel{.}{=} 1/T$
(c) the odd-order harmonic components $3f$, $5f$, $7f$, $9f$, etc.

The amplitudes of the harmonic components are smaller than the amplitude of the fundamental by the same ratio as their order. If the amplitude of the fundamental component is V, then the amplitude of

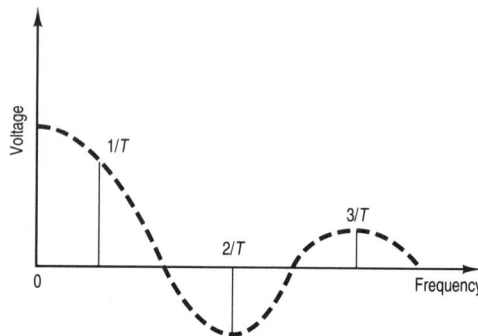

Fig. 9.11 Spectrum diagram of a square wave.

the third-harmonic component is $V/3$, the amplitude of the fifth-harmonic component is $V/5$, and so on.

A triangular wave has exactly the same harmonic content as a square wave but the amplitudes of the harmonics are different. If, again, the amplitude of the fundamental-frequency component is V, the amplitudes of the third and the fifth harmonics are $V/9$ and $V/25$ respectively. This means that the amplitude of each harmonic decreases with the square of the order of the harmonic.

Lastly, a sawtooth waveform contains components at a fundamental frequency, plus both even and odd harmonics.

It is not possible to transmit all the frequency components of a rectangular, sawtooth, or triangular waveform over a transmission system because the required bandwidth would be too wide. In practice, therefore, not all the component frequencies of such waveforms are transmitted. This means that the signal waveform received at the destination will not be identical to the original waveform and hence some distortion will have occurred. The wider the bandwidth of a system the greater will be the number of harmonics that can be transmitted over that system, but the price to be paid for this is increased costs. In practice, therefore, the bandwidth provided for a system is the smallest possible that will still give adequate reproduction of the transmitted waveform.

Speech transmission

In a telecommunication system speech information is transmitted from one point to another; telephone conversations are sent over telephone cables and radio links and it is always necessary to make sure that enough information arrives at the receiving end to allow the intelligence to be understood. For speech to be understood, it is necessary to ensure that the received signal contains most of the frequency components that were in the original speech waveform and also to ensure that as few as possible are lost during transmission.

When a person speaks the sound waves produced contain frequencies ranging from about 100 Hz to about 10 kHz. The pitch of the sound wave is determined by its fundamental frequency and this is typically about 100 to 500 Hz for men and 200 to 1000 Hz for women. The human ear is able to respond over a frequency range of about 30 Hz to 16 kHz and is most sensitive in the frequency band 1 to 2 kHz. However, it is not economically possible to transmit such relatively wide bandwidths. It has been found that a much smaller bandwidth will give satisfactory results and so the internationally agreed bandwidth for speech is 300 to 3400 Hz. The use of this 'commercial quality speech' bandwidth means that neither the highest nor the lowest frequencies in a speech waveform are transmitted. However, to the human ear the pitch of a sound is set by the frequency difference between the harmonics in the wave and so, even though the fundamental frequency may have been suppressed, the pitch of the received sound is the same as in the original speech. The loss of frequencies above 3400 Hz reduces the quality of the speech but does not adversely affect its intelligibility.

Telephone transmitter

Speech information is converted into the corresponding analogue electrical signal by means of a transducer known as a **microphone**. There are several different types of microphone in common use but the easiest to understand is the **carbon granule** type that is used in older types of telephone instrument. The carbon granule microphone is often known as a **telephone transmitter**.

The basic construction of a carbon granule telephone transmitter is shown in Fig. 9.12. A chamber, into which carbon granules are packed, has electrodes mounted at either side to which the external circuitry is connected. One of the electrodes is fixed in position while the other electrode can have its position varied by the movement of a diaphragm. When the diaphragm is moved towards the chamber, the carbon granules are pushed more closely together and this action reduces their contact resistance, i.e. the resistance between two touching granules. The closer granule packing results in the resistance between the two electrodes becoming smaller. Conversely, when the diaphragm is moved out from the chamber the carbon granules are then less tightly packed together and so their contact resistance increases. The total resistance between the two electrodes then increases also. If, therefore, a constant voltage is maintained across the transmitter the effect of moving the diaphragm in and out is to produce a varying current flow through the device.

A sound wave consists of variations in atmospheric pressure either side of the normal value. If a sound wave is incident upon a telephone transmitter the changes in pressure will cause the transmitter's diaphragm to vibrate and this vibration, in turn, will cause the current flowing through the transmitter to vary with the same waveform as the incident speech. [This is not strictly true because the telephone transmitter introduces some distortion of the signal waveform.] The variation of the current flowing through a telephone transmitter is illustrated in Fig. 9.13.

If the telephone transmitter were to be connected directly in series with the telephone line the variations in its resistance caused by a sound wave incident upon its diaphragm could well be swamped by the resistance of the line. In a telephone instrument, therefore, the transmitter is isolated from the line by a transformer, as shown in Fig. 9.14. The analogue electrical signal that is transmitted to line is known as the **baseband** signal.

Modern telephone instruments employ electronic circuitry using ICs and incorporate multi-frequency signalling and a keypad. They also use an Electret transmitter instead of a carbon granule transmitter. An Electret transmitter uses two electrically charged plates that are

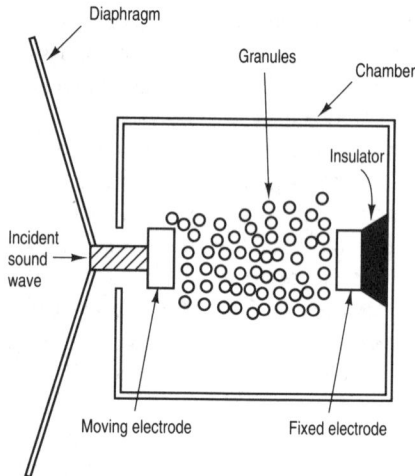

Fig. 9.12 Carbon granule telephone transmitter.

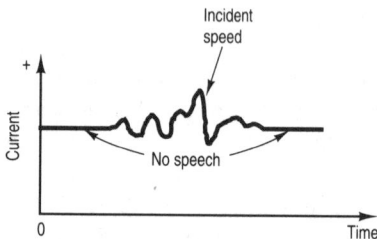

Fig. 9.13 Current flowing in a carbon granule transmitter.

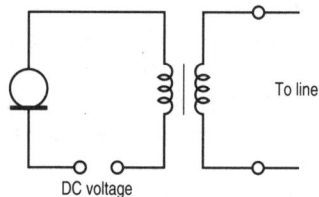

Fig. 9.14 Isolation of a telephone transmitter from line.

caused to vibrate by an incident sound wave. The vibrations cause a small voltage to be generated and this voltage is amplified by an IC. This type of transmitter gives better quality reproduction of a speech waveform and also better reliability.

Telephone receiver

At the receiving end of a telephone connection the received analogue electrical signal must be converted back into the corresponding sound wave and this conversion is the function of another transducer. This transducer is known as a **telephone receiver** and its basic construction is shown in Fig. 9.15.

A permanent bar magnet has a yoke and two pole pieces connected to it, and each pole piece is wound with a coil of wire. The armature is mounted on a pivot at one end of the magnet. The magnetic force of attraction between a north and a south pole will ensure that the armature will be held against one or other of the two pole pieces. If a current is passed through the coils then, according to its direction of flow, the magnetic field in one of the air gaps will be increased and the magnetic field in the other air gap will be reduced. The armature will then be attracted to the pole piece where there is the stronger magnetic field. If the direction of the current flowing through the coils is then reversed the magnetic fields set up by the current will also reverse their direction. The strength of the magnetic fields in the two air gaps now changes, one field is strengthened while the other is weakened. This causes the armature to change over and rest against the other pole piece. If an a.c. current is passed through the coils the armature will continually move from one pole piece to the other and this movement of the armature is translated, via the driving pin, to the diaphragm. The vibrations of the diaphragm then generate a sound wave which is a reasonable reproduction of the original speech waveform at the other end of the telephone connection.

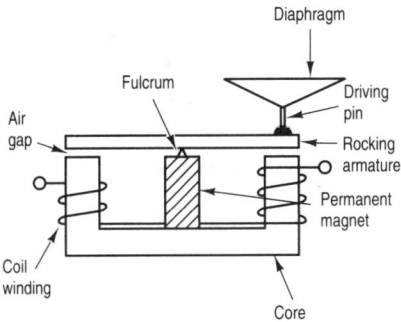

Fig. 9.15 Telephone receiver.

Modulation

Modulation is the process of superimposing information onto a carrier wave. In the process one of the characteristics of the carrier is varied by the modulating or **baseband**, signal.

For economic reasons telephone circuits, other than local lines, are nearly always routed over multi-channel telephony systems which make it possible to transmit many circuits over a single pair of conductors in a cable. The technique that is employed in the UK telephone network is known as **time-division multiplex** and it uses a modulation method known as **pulse code modulation** (PCM). Data signals cannot be directly transmitted over analogue telephone circuits, for reasons discussed in Chapter 11, and some form of digital modulation is necessary to change the signals into **voice-frequency form**. The voice-frequency signal can be transmitted over the

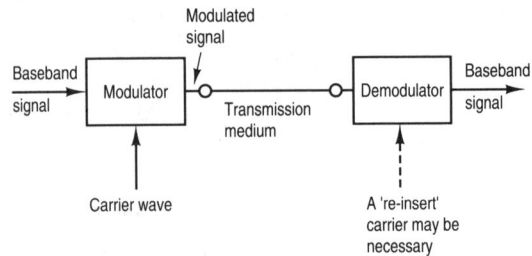

Fig. 9.16 Principle of modulation.

telephone network in exactly the same way as speech signals. Lastly, whenever a signal is to be transmitted over a radio link it must first be shifted to some higher frequency band. There are two reasons why this **frequency translation** is necessary: (a) aerials will not work at frequencies below about 15 kHz; and (b) it is necessary for different signals to be transmitted at different frequencies so that the radio receiver can select a particular signal from all the signals picked up by the aerial.

The basic principle of modulation is shown in Fig. 9.16. An audio baseband signal in the frequency band 0 to 10 kHz is applied to a **modulator** along with a 10 MHz carrier wave. The action of the modulator is to modulate the carrier with the baseband signal. The output of the modulator is a modulated wave that occupies a frequency band centred on 10 MHz and this modulated wave is applied to a transmitting aerial. The signal is propagated through the atmosphere and is picked up by the receiving aerial. The desired signal is selected from all the other signals that are also picked up by the aerial, and is amplified and applied to a **demodulator** (or detector) which recovers the baseband signal from the modulated wave. Demodulation is the reverse process to modulation.

Amplitude modulation

With amplitude modulation the **amplitude** of a sinusoidal carrier wave is varied by the modulating, or baseband, signal. The amplitude of the variations is equal to the baseband signal voltage and the number of times per second the variation occurs is equal to the baseband frequency. The frequency of the carrier remains unchanged.

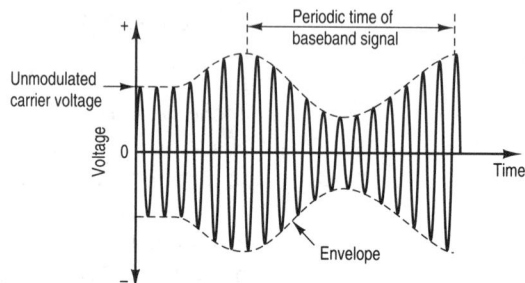

Fig. 9.17 Amplitude modulation.

Amplitude modulation of a sinusoidal carrier wave by a sinusoidal modulating signal is shown in Fig. 9.17. The frequency f_c of the carrier wave is usually several times higher than the baseband (modulating) frequency f_m. The amplitude of the carrier can be seen to increase when the baseband signal increases in the positive direction, and reaches its maximum value when the baseband signal is at its maximum positive value. Conversely, during the alternate half-cycles of the baseband signal the carrier amplitude decreases as the baseband voltage increases in the negative direction, and it reaches its minimum value when the baseband voltage is at its maximum negative value. The outline of the modulated carrier wave, which is shown dotted, is known as the **envelope** of the wave. The envelope always has the same waveform as the baseband signal.

A sinusoidally modulated carrier wave contains components at three different frequencies. These are:

(a) the original carrier frequency, f_c;
(b) the sum of the carrier frequency and the baseband frequency, known as the **upper sidefrequency**, $f_c + f_m$;
(c) the difference between the carrier frequency and the baseband frequency, known as the **lower sidefrequency**, $f_c - f_m$.

The baseband frequency f_m is *not* present in the modulated wave. The frequencies of the input and output signals of an amplitude modulator are shown in Fig. 9.18. The bandwidth occupied by the modulated wave is equal to the difference between the highest frequency in the wave and the lowest frequency. Therefore,

$$\text{bandwidth} = (f_c + f_m) - (f_c - f_m) = 2f_m \qquad (9.3)$$

and is equal to twice the frequency of the baseband signal.

Baseband signal f_m → Amplitude modulator → Amplitude-modulated wave f_c, $f_c \pm f_m$

Carrier wave f_c

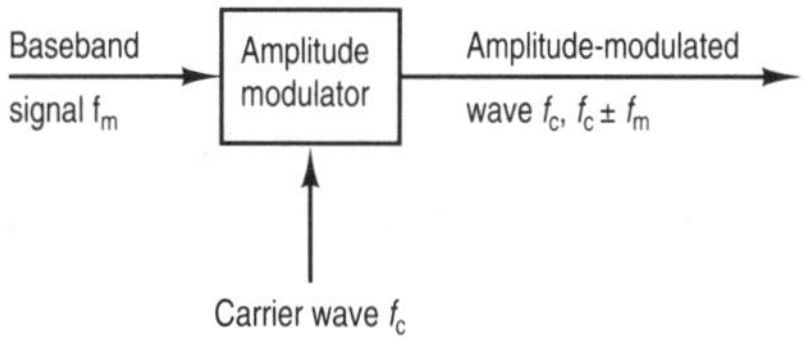

Fig. 9.18 Input and output frequencies of an amplitude modulator.

Example 9.5

A 80 kHz carrier wave is amplitude modulated by a sinusoidal signal of frequency 3 kHz. Determine the frequencies contained in the modulated wave and the bandwidth needed for its transmission.

Solution
The frequencies contained in the modulated wave are:

(a) The carrier frequency $f_c = 80$ kHz. (*Ans.*)
(b) The lower sidefrequency $f_c - f_m = 80$ kHz $- 3$ kHz
    $= 77$ kHz. (*Ans.*)
(c) The upper sidefrequency $f_c + f_m = 80$ kHz $+ 3$ kHz
    $= 83$ kHz. (*Ans.*)

The required bandwidth $= 6$ kHz. (*Ans.*)

The two sidefrequencies are symmetrically situated on each side of the carrier frequency and separated from it by a frequency gap

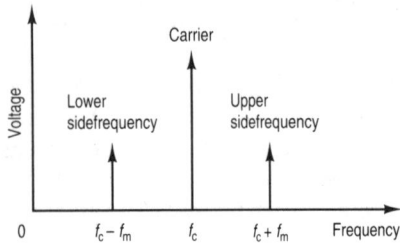

Fig. 9.19 Spectrum diagram of a sinusoidally amplitude-modulated wave.

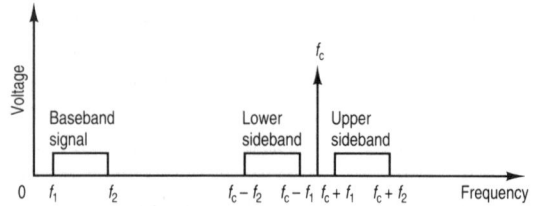

Fig. 9.20 Spectrum diagram of a non-sinusoidally amplitude-modulated wave.

equal to the baseband frequency f_m. This is shown by the frequency spectrum diagram given in Fig. 9.19.

In practice, of course, the baseband signal would very rarely be of sinusoidal waveform but would be a non-sinusoidal signal occupying a certain bandwidth. If the baseband signal occupies the bandwidth from frequency f_1 to frequency f_2 then the frequency spectrum diagram will be as shown in Fig. 9.20. Two **sidebands** are produced, the lower sideband contains frequencies from $f_c - f_2$ to $f_c - f_1$ and the upper sideband contains frequencies in the band $f_c + f_1$ to $f_c + f_2$. The minimum bandwidth required is equal to the highest frequency contained in the modulated wave minus the lowest frequency, i.e.

$$(f_c + f_2) - (f_c - f_2) = 2f_2 \tag{9.4}$$

This means that the minimum bandwidth is equal to twice the highest frequency contained in the baseband signal.

Example 9.5

A 700 kHz carrier wave is amplitude modulated by the band of frequencies 300 to 3400 kHz. Calculate the frequencies that are contained in the modulated wave and the necessary bandwidth to transmit the signal.

Solution
The frequencies contained in the modulated wave are:

(a) The carrier frequency 700 kHz. (*Ans.*)
(b) In the lower sideband: from $(700 - 3.4)$ kHz to $(700 - 0.3)$ kHz $= 696.6$ to 699.7 kHz. (*Ans.*)
(c) In the upper sideband: from 700.3 to 703.4 kHz. (*Ans.*)

Frequency modulation

When a sinusoidal carrier wave is frequency modulated (FM) the frequency of the carrier is varied by the baseband signal and the carrier voltage is kept at a constant value. The amount by which the carrier

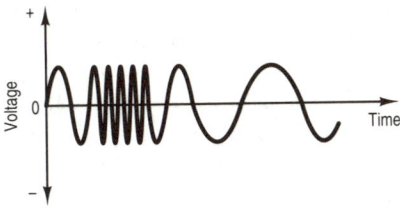

Fig. 9.21 Frequency-modulated wave.

frequency is deviated from its unmodulated value is directly proportional to the baseband voltage and the number of times per second the deviations occur is equal to the baseband frequency. Figure 9.21 shows a frequency-modulated wave. When the baseband signal voltage is increasing in the positive direction the carrier frequency is also increasing. Maximum frequency deviation occurs when the baseband voltage has reached its maximum positive value. As the baseband voltage decreases from its maximum positive value towards zero volts the carrier frequency falls and reaches its unmodulated value when the baseband voltage is instantaneously zero. When, now, the baseband voltage becomes negative the carrier frequency reduces below its unmodulated value and reaches its minimum value when the baseband voltage is at its maximum negative value. Note that the carrier voltage remains constant.

When a carrier is frequency modulated by a sinusoidal signal the modulation process may produce a number of components at the following frequencies:

$$f_c, f_c + f_m, f_c + 2f_m, f_c + 3f_m, f_c + 4f_m, \text{ and so on.}$$

It can be seen that a number of orders of sidefrequencies are generated; all of these components, including the carrier, have amplitudes that vary with the baseband voltage and any of them may, again including the carrier, be equal to zero. As the spectrum diagram for a frequency-modulated waveform is complex, it is not given here. Because a large number of components at different frequencies are generated, even for sinusoidal modulation, it is not possible to transmit all of them and an arbitrary bandwidth must be assigned. A frequency-modulated radio system may be **narrow-band** when it occupies more or less the same bandwidth as an amplitude-modulated signal, or it may be **wideband** when the occupied bandwidth is much larger. Narrow-band FM systems are used for mobile radio and wideband FM systems are used for both sound broadcasting and television sound.

Pulse modulation

It is also possible to modulate a repetitive pulse waveform carrier instead of a sinusoidal carrier to obtain pulse modulation. Four main types of pulse modulation exist: **pulse amplitude modulation, pulse duration modulation, pulse position modulation**, and **pulse code modulation**. The first three techniques were often used in the past but now their applications are severely limited; however, pulse amplitude modulation is used as a first step in the production of a pulse-code-modulated signal and pulse duration modulation is employed in some audio power amplifiers. The basic concepts of the first three modulation systems are illustrated by the waveform diagrams given in Fig. 9.22.

With pulse amplitude modulation, Fig. 9.22(a), pulses of equal

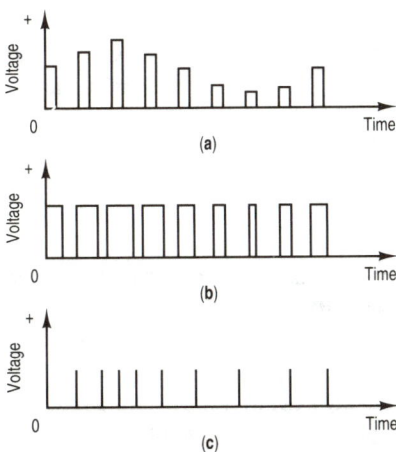

Fig. 9.22 Pulse modulation: (a) pulse amplitude modulation; (b) pulse duration modulation; and (c) pulse position modulation.

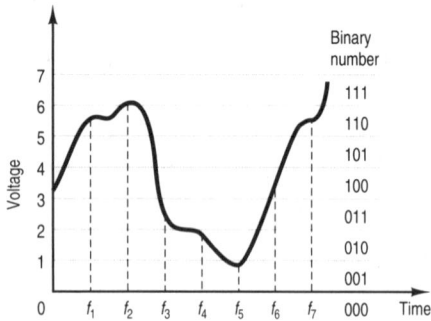

Fig. 9.23 Principle of pulse code modulation.

width and spacing have their amplitude modulated by the baseband signal. The other two systems employ pulses of constant amplitude; a pulse-duration-modulated signal, Fig. 9.22(b), has *either* the leading edge, or the trailing edge of each pulse modulated in time. In the figure the leading edges are in constant positions and the trailing edges are modulated. A pulse-position-modulated wave, Fig. 9.22(c), consists of a series of very narrow pulses whose positions in time have been modulated; these positions correspond with the positions of the modulated pulse edges of the pulse-duration-modulated waveform.

The fourth pulse-modulated system, pulse code modulation (PCM), is widely employed in modern telecommunication networks to provide multi-channel telephony systems. In a PCM system the signal to be transmitted is first converted into a pulse-amplitude-modulated waveform, then the height of each pulse is measured and a binary coded digital waveform is transmitted to represent that voltage. To reduce the number of possible amplitudes that need to be represented the voltage range to be transmitted is divided into a number of **quantization levels**. In the systems used in the UK telephone network 256 levels are employed but, to simplify the drawing, only 8 are shown in Fig. 9.23.

Digital modulation

Digital data signals cannot be transmitted over the analogue transmission network for reasons discussed in Chapter 11, and so they must be converted into voice-frequency signals using some form of digital modulation. Digital signals can be represented by:

(a) two different amplitudes in an amplitude-modulated system;
(b) two different frequencies in a frequency-modulated system;
(c) two different phases in a phase-modulated system.

For various reasons digital amplitude modulation on its own is rarely employed although it does get used in conjunction with phase modulation at higher bit rates — in a system known as **quadrature amplitude modulation**.

Frequency shift keying

For historical reasons frequency modulation of a digital data signal is generally known as **frequency shift keying** (FSK). When a data signal is applied to a FSK modulator the frequency of a sinusoidal carrier wave is switched between two different values at the bit rate of the signal. The higher frequency is used to represent binary 0 and the lower frequency represents binary 1. For example, in the V23 FSK system the frequencies used are 1300 Hz and either 1700 Hz or 2100 Hz. Figure 9.24 shows a digital data signal and the corresponding FSK waveform. The minimum bandwidth required for the transmission of an FSK signal is equal to the bit rate. Because the

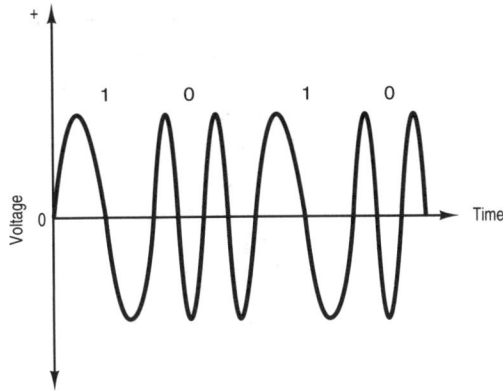

Fig. 9.24 PSK signal.

necessary separation between the two frequencies is a function of the bit rate, and because the bandwidth made available by the telephone network is limited, FSK is only used for bit rates of 600 and 1200 bits/s.

Phase shift keying

The phase shift keying systems employed in data communication use **changes** in the phase of a sinusoidal carrier wave to represent combinations of bits; before the modulation process is carried out the bits are grouped together either in twos — known as **dibits** — or in threes — known as **tribits**. The modulation technique that is used is known as **differential phase shift keying** (DPSK). The use of dibits or tribits reduces the bandwidth that must be provided to transmit a DPSK signal and, for example, a 2400 bits/s DPSK signal occupies a bandwidth of only 1200 Hz.

Quadrature amplitude modulation

Quadrature amplitude modulation (QAM) — a mixture of amplitude modulation and phase modulation — is employed at bit rates of 1200 bits/s and higher. Bits are grouped into fours, or **quabits**, before the data signal modulates the carrier. The 16 possible combinations of quabits can each be indicated by choosing any one of either (a) two amplitudes and eight phases or (b) four amplitudes and four phases.

Noise

All electronic, computer, and telecommunication systems are subject to interference from electrical noise. Any system must be designed to minimize the adverse effects of such noise and this means that (a) the generation of man-made noise voltages must be reduced as much as possible, (b) circuits should be screened, or shielded, from interfering voltages, and (c) the wanted signal voltage level must always be kept several times larger than the unwanted noise voltages. The last requirement is expressed in terms of a relationship that is known as the signal-to-noise ratio:

$$\text{Signal-to-noise ratio} = \frac{\text{(wanted signal power)}}{\text{(unwanted noise power)}} \quad (9.4)$$

Example 9.6

The signal power received at the end of a communication circuit is 1 mW. Calculate the signal-to-noise ratio at this point if the noise level is 1 μW.

Solution
Signal-to-noise ratio = $(1 \times 10^{-3})/(1 \times 10^{-6})$ = 1000. (*Ans.*)

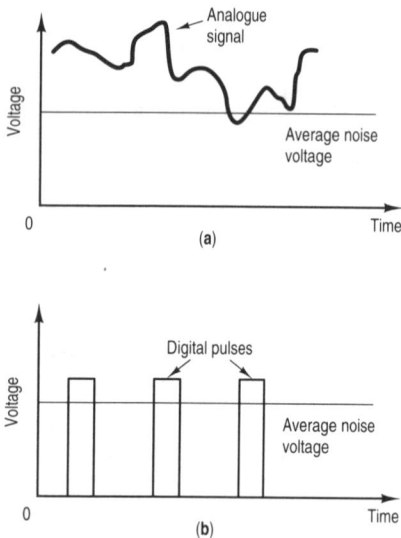

The minimum allowable figure for signal-to-noise ratio varies with the type of system considered, but it is always a very important consideration. Analogue systems are always affected by noise to some extent but a digital system can be made more or less immune to noise provided the signal-to-noise ratio is kept higher than some threshold value. This point is illustrated in Fig. 9.25. In figure (a) the analogue voltage is sometimes smaller than the average noise voltage, and at other times is much larger. This would result in a very poor signal-to-noise ratio. In figure (b) the digital pulses are kept above the average noise level. The receiver is always able to decide whether or not a pulse is present and a good signal-to-noise ratio is obtained.

The main types of electrical noise are as follows:

Fig. 9.25 Effect of noise in (a) an analogue system and (b) a digital system.

1. **Resistor noise**. Noise voltages are always generated in resistors and, indeed, in the resistance of any electrical conductor.
2. **Semiconductor noise**. Noise voltages and currents are always generated using several mechanisms within a semiconductor device, such as a transistor, a diode, or an IC.
3. **Man-made noise**. Interfering noise voltages are likely to be generated in any electrical equipment in which an electrical current is abruptly switched on or off. The possible sources of such noise are many and include electric light switches, brushes on electric motors in domestic appliances such as vacuum cleaners and hair driers, motor car ignition systems, and neon lights.
4. **Natural noise**. Noise is produced by thunderstorms, by the sun, and by the distant stars.
5. **Crosstalk**. Crosstalk occurs between nearby conductors, between which there exists unwanted capacitive and inductive couplings. Telephone cables with copper conductors are constructed, and electronic circuits are laid out, in ways that reduce these unwanted couplings to the minimum possible. Optical fibre cables do not suffer from crosstalk problems.

10 The public switched telephone network

The user of a telephone at home or in an office expects to be able to communicate with the user of any other telephone that is also connected to the telephone network. It is not possible to directly connect every telephone in a given area to every other telephone in that area, let alone to every other telephone throughout the country. Hence some form of switched network is essential. Every telephone is connected, via the **access network**, to a local telephone exchange that is able to interconnect any two of the lines coming into the exchange. Very often a caller will want to communicate with a number on some other telephone exchange and so it is also necessary for the local telephone exchanges to be interconnected. It is not economic for all the local exchanges to be fully interconnected and hence only those exchanges that have sufficient telephone traffic between them to justify the expense are provided with direct links. All other local exchanges are linked together when necessary by trunk-switching exchanges.

The **public switched telephone network** (PSTN) of the United Kingdom now consists mainly of digital telephone exchanges (about 85% in 1993) that are linked together by digital transmission systems that employ time division multiplex (TDM) and pulse code modulation (PCM). The basic block diagram of the telephone network is given in Fig. 10.1. The transmission media that carry the PCM systems include coaxial cable, optical fibre cables, and microwave radio relay systems, with optical fibre cables now carrying most of the traffic. The relative merits of the three different transmission media are discussed later. The access network is still predominantly provided by audio star quad and multiple twin cables carrying analogue signals,

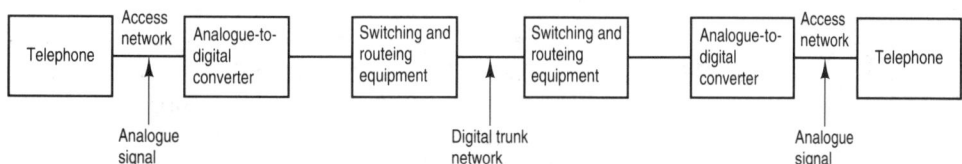

Fig. 10.1 A connection in the digital telephone network.

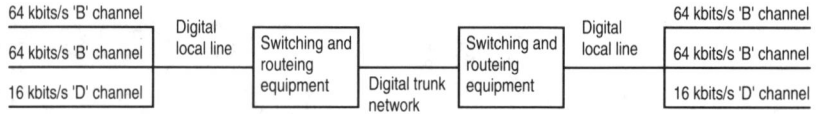

64 kbits/s 'B' channel	Digital local line	Switching and routeing equipment	Digital trunk network	Switching and routeing equipment	Digital local line	64 kbits/s 'B' channel
64 kbits/s 'B' channel						64 kbits/s 'B' channel
16 kbits/s 'D' channel						16 kbits/s 'D' channel

Fig. 10.2 Digital access to the digital telephone network.

although eventually this network will be provided by optical fibre cables carrying digital signals. The local ends will then provide two 64 kbits/s 'B' channels for the transmission of either data or speech signals and one 16 kbits/s 'D' channel for signalling and control of the 'B' channel. The block diagram of the digital access circuit is shown in Fig. 10.2.

Telephone circuits

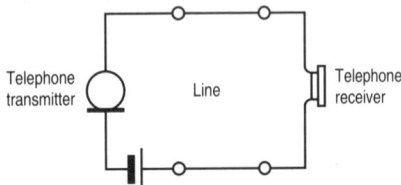

Fig. 10.3 Unidirectional telephone circuit.

The local battery telephone circuit

To obtain a simple unidirectional telephone circuit a telephone transmitter must be connected to a telephone receiver by a length of transmission line. The basic arrangement of a simple telephone circuit is shown in Fig. 10.3. When the telephone transmitter is spoken into it causes a speech-frequency current to flow around the circuit. As this current passes through the telephone receiver at the other end of the circuit the receiver converts the electrical signal into a sound signal. The reproduced sound is the same as the original speech wave that was incident upon the distant telephone transmitter, plus some distortion caused by the non-perfect actions of both the transmitter and the receiver. This simple telephone circuit, however, has two main disadvantages:

1. Another identical circuit will be necessary if it is also required to transmit speech in the other direction.
2. The longer the length of line, the smaller will be the changes in the line current and the received speech may be very faint.

To overcome these difficulties the slightly more complex arrangement shown in Fig. 10.4 may be used. In this circuit a local battery is provided at each telephone to power the telephone transmitter, and the battery/transmitter circuit is isolated from the telephone line by a transformer. Now an incident speech signal causes the telephone transmitter to produce relatively large current variations in the transmitter loop. This changing current induces an alternating e.m.f. into the secondary winding of the transformer and this voltage, in turn, causes a current, having the same waveform as the speech signal (assuming zero distortion), to flow into the telephone line. At the other end of the line this current flows through the telephone receiver and the receiver then reproduces the original speech signal. This kind of circuit was common in the early days of telephony but it suffers from the big disadvantage of needing a separate battery at each telephone. To overcome this snag the power required to operate each telephone can be obtained from the local telephone exchange.

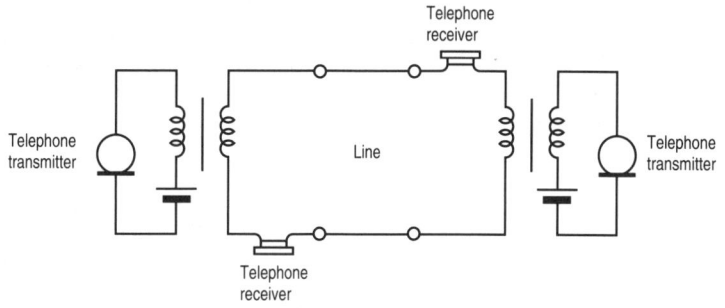

Fig. 10.4 Local battery telephone circuit.

The central battery telephone circuit

The basic telephone circuit in a **central battery** system is shown in Fig. 10.5. A large-capacity-50 V battery, installed at the telephone exchange, supplies d.c. current to each telephone connected to the exchange when the handset is picked up. When any telephone is spoken into, the d.c. current flowing through it is made to vary, with the same waveform as the speech, and this changing current flows over the local line to the telephone exchange. The changing current effectively consists of the d.c. battery current with a speech-frequency a.c. current superimposed upon it. At the telephone exchange a large-value inductance connected in series with the exchange battery prevents the speech-frequency component of the changing current from passing into the exchange battery. Instead, the a.c. component of the current flows down the other telephone line and through the distant telephone receiver and this, in turn, reproduces the original speech. As it stands this simple circuit would suffer very badly from an effect known as **sidetone**, i.e. any sounds picked up by a telephone transmitter will be heard in the associated telephone transmitter. The sidetone arises, of course, because the speech-frequency current produced by a transmitter also flows through its associated receiver. Sidetone makes it very difficult, if not impossible, to carry on a conversation and so steps must be taken to reduce it. Sidetone should not, however, be completely eliminated otherwise the telephone would appear to be 'dead' to a user.

The basic circuit of a dial-type telephone instrument is shown in

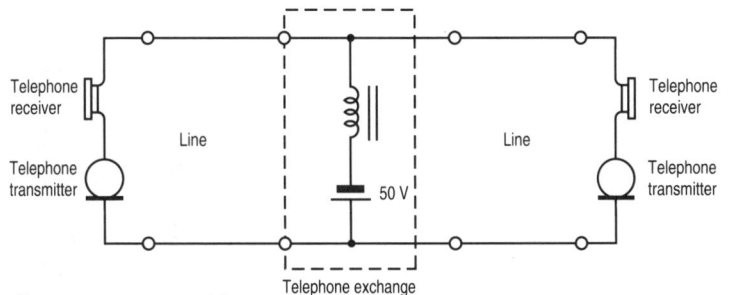

Fig. 10.5 Central battery telephone circuit.

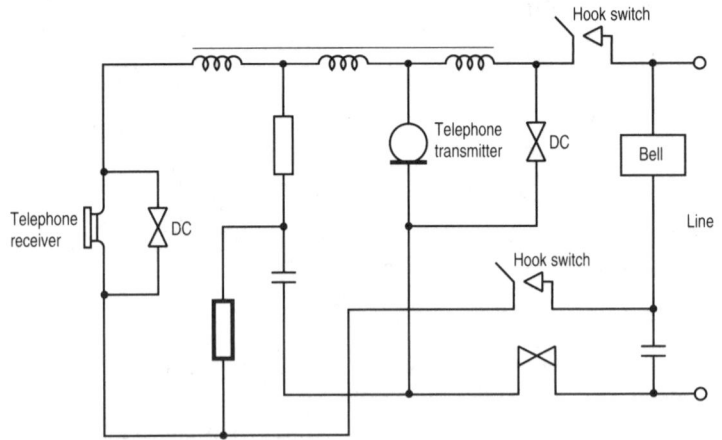

Fig. 10.6 Dial-type telephone.

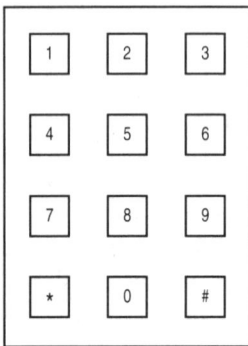

Fig. 10.7 Standard telephone keypad.

Fig. 10.6. When the handset is lifted from the instrument the contacts of the hook switch make and complete a path for the d.c. current supplied by the local exchange to flow via the telephone transmitter. The dial gives **loop-disconnect** signalling between the telephone and the local telephone exchange; when the dial is operated its contacts break the line loop a number of times equal to the number dialled (0 = ten). These breaks are interpreted by the exchange equipment to route the call to its wanted destination. The contacts marked DC, connected across both the receiver and the transmitter, prevent dial impulses passing through these transducers and so prevent the pulses being audible. Sidetone is reduced to a low level by the tapped transformer arrangement.

Modern telephones use keypad 'dialling' and multi-frequency signalling, and all the internal circuitry is provided by one or more ICs. Facilities like 'number recall' and 'last number re-dial' are provided by a microcontroller. The standard keypad is shown in Fig. 10.7. When a key is pressed, conducting rubber is pressed on to a printed circuit board to place a low-resistance contact between two points in the telephone circuit. Each such contact is detected by the electronic circuitry and two tones are sent to line to indicate the particular number 'dialled'. Each dual-tone signal is transmitted to the local exchange and here it is decoded into the digits it represents. The frequencies employed are given in Table 10.1.

Table 10.1

Digit	Tones	(Hz)		Digit	Tones	(Hz)
1	697	1209		7	852	1209
2	697	1336		8	852	1336
3	697	1477		9	852	1477
4	770	1209		0	941	1209
5	770	1336		*	941	1336
6	770	1477		#	941	1477

Telephone exchange switching

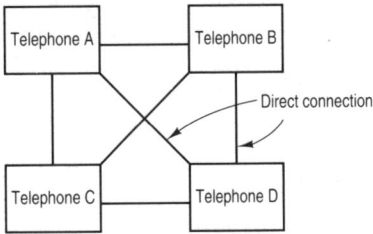

Fig. 10.8 4-Telephone network.

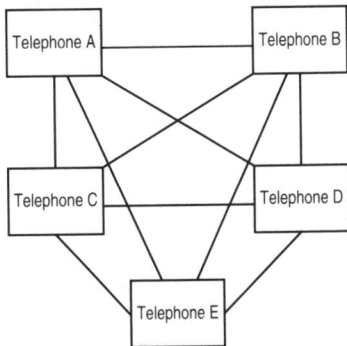

Fig. 10.9 5-Telephone network.

Every telephone connected to a telephone network ought to have access to every other telephone connected to that network. If the network is small, like the 4-telephone network shown in Fig. 10.8, it is quite possible to provide a direct link between each pair of telephones. Note that the number of direct connections between the telephones is 6. If the number of telephones in the network is increased to 5, as in Fig. 10.9, then the number of direct connections required increases to 10. Similarly, if the number of telephones is 6 the number of direct connection necessary becomes equal to 15, and for 7 telephones equal to 21. In general, if there are n telephones then $n(n-1)/2$ direct connections are required to obtain a fully interconnected network. Clearly as the number of telephones in a network increases the number of direct circuits required to give a fully interconnected network rapidly becomes impossibly high and therefore some form of switching is necessary.

The basic concept of a switched telephone system is shown in Fig. 10.10; eight telephones are each directly connected to a switching centre that is able to switch any of the telephones to any other telephone. In a small office the switching centre may be either a small unit and an operator's console, or a private branch automatic exchange (PABX). In the public telephone system the switching centre would be the local telephone exchange. The majority of the local telephone exchanges in the BT network are common-control digital exchanges. Common control means that the setting up of a call through a telephone exchange is controlled by a central processor, and this means that all the 'dialled' digits must be received before the setting up of the call path can begin. The method of setting up a connection is as follows: a caller indicates to the exchange that he or she wishes to make a call by lifting the handset from its rest. This action completes a d.c. path and allows the exchange battery to supply a d.c. current to the calling telephone. The completion of a call is simply indicated by the caller replacing the handset on its rest to break the d.c. loop. The exchange now identifies the number of the calling telephone and allocates equipment that will be able to set up and control the required connection. Dialling tone is now returned to let the caller know that

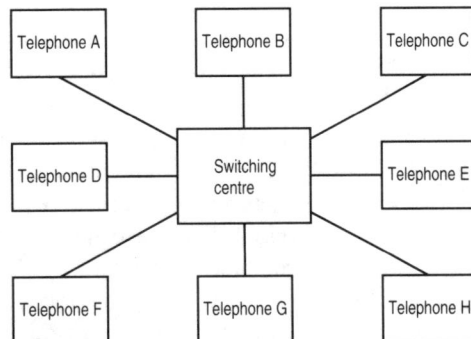

Fig. 10.10 Switched telephone network.

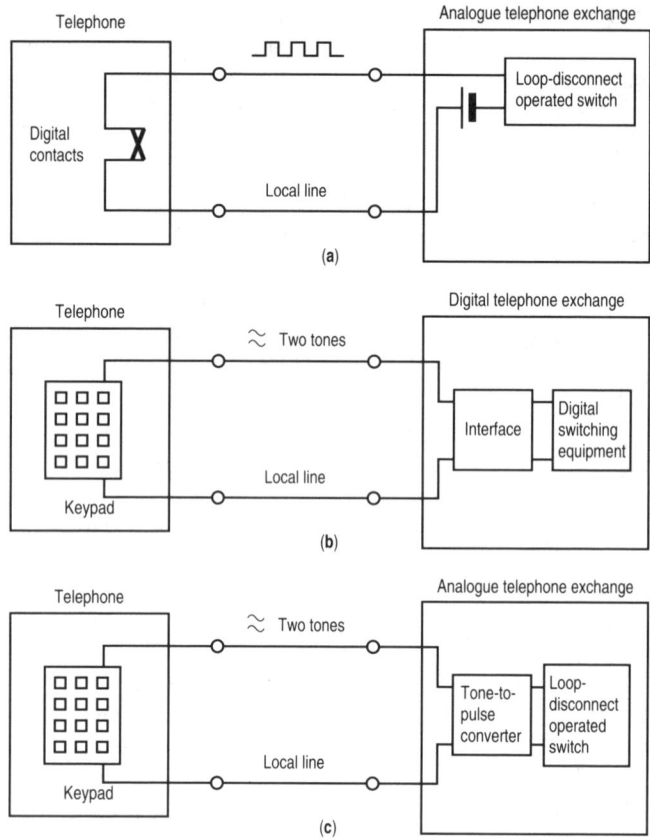

Fig. 10.11 (a) Loop-disconnect signalling; (b) DTMF signalling to a digital exchange; (c) DTMF signalling to an analogue exchange.

'dialling' may now begin. Note that the term 'dialling' is still employed even when a keypad telephone is used.

The dialled digits are sent to the exchange using either **loop-disconnect**, or **dual-tone multi-frequency** (DTMF), signalling. With loop-disconnect signalling (see Fig. 10.11(a)), the d.c. path between the telephone and the local exchange is broken a number of times by a device known as a **dial**, the number of breaks corresponds to the number dialled, with 0 = 10 breaks. Loop-disconnect signalling has been used for many years but it has a number of disadvantages: (a) it is slow, (b) it suffers from severe distortion over long local lines, and (c) it requires a d.c. path. The more modern DTMF method of signalling is shown by Figs 10.11(b) and (c). In a digital exchange the necessary conversion is carried out by the customer's line interface unit but in an older analogue exchange a tone-to-pulse converter is required to convert the tones into the loop-disconnect signal that the exchange equipment needs. If a loop-disconnect dialled number is sent into a digital exchange the interface circuit can convert it into the required digital codes. DTMF signalling is more convenient, more efficient, of greater reliability, and can be extended to the remote control and/or interrogation of answering machines and computers.

If a call is made to another number on the same telephone exchange the dialled digits directly indicate the outgoing (local) line to be selected. If the called line is engaged, this fact is signalled back to the caller by the return of the **engaged tone**. Should the called line be free, the exchange applies ringing current to the called line and returns ringing tone to the caller and the connection is established through the exchange. As soon as the user of the called telephone answers the call, the exchange removes both the ringing current and the ringing tone and the conversation may then begin. The connection is then supervised by the exchange awaiting the clear-down signal initiated by a handset being replaced on its rest; when this occurs the connection is cleared down and both lines are returned to their free state. Both lines are then free to initiate another call. The exchange also registers the duration of the call so that it can later be charged.

Digital telephone exchanges

The basic block diagram of a digital telephone exchange is shown in Fig. 10.12. Each telephone is connected to the exchange by a line in the access network.

BORSCHT

The local line terminates on an interface circuit that has a number of functions and that is often known as a BORSCHT, a label that stands for battery, overvoltage, ringing, signalling, coding, hybrid, and testing. The main functions of this interface circuit are as follows:

1. To supply d.c. current to each telephone when its handset is lifted from its rest. This is necessary for two reasons: (a) so that a calling condition can be detected, and (b) so that power can be supplied to energize the telephone transmitter and the telephone's internal electronic circuitry.
2. To protect against any high voltages that may be picked up by the local line.
3. To apply ringing current to a called line. In the UK ringing current is supplied at 20 Hz at 2 second intervals ON and 4 second intervals OFF.

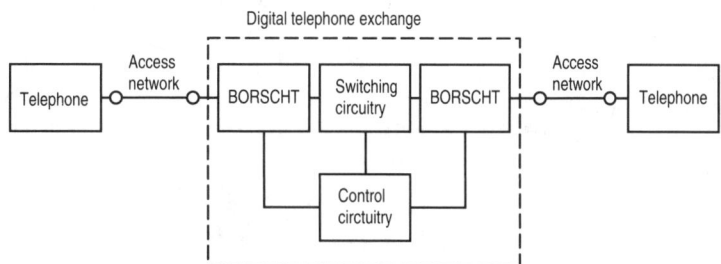

Fig. 10.12 Principle of a digital telephone exchange.

4. To detect when a handset is lifted (known as the 'off-hook condition') and when it has been replaced (known as the 'on-hook' condition). Also to detect 'dialled' numbers whether they employ loop-disconnect or DTMF signalling.

5. To convert all analogue lines into digital operation. This requirement is satisfied by a circuit known as a **codec** that converts the analogue speech signals into PCM, or, in the other direction of transmission, converts a PCM signal into an analogue speech signal.

6. To convert the two-wire local line in the access network into the four-wire circuit that will be switched by the exchange equipment.

7. Some testing facilities are also provided to determine whether a line is free or busy.

Control circuitry

The control circuitry uses **stored program control** (SPC) to provide all the control and supervision necessary to establish, monitor, and clear a call through the exchange. The circuitry is software controlled and may easily be modified to suit changing requirements.

Switching circuits

The switching circuitry must be able to connect any one of its inputs to any one of its outputs. This means that every local line connected to a telephone exchange must be able to be connected to every other local line or to any outgoing trunk circuit to another exchange. The basic switching matrix is shown in Fig. 10.13(a). The matrix has four inlets and four outlets. The intersections between the inlets and the outlets are called **crosspoints**. At each crosspoint an electronic switch is provided that is able to make a connection between an inlet and an outlet and this enables any one of the four inlets to be connected to any one of the four outlets. Two examples are shown in Figs 10.13(b) and (c) respectively; in figure (b) inlet 2 is connected to outlet 1 and inlet 3 is connected to outlet 3. In figure (c) the switch is fully loaded, i.e. all four inlets are in use; inlet 1 is connected to outlet 3, 2 to 2, 3 to 1, and 4 to 4. The number of inlets and outlets are not necessarily equal but the number of crosspoints is always equal to the product of the number of inlets times the number of outlets. The maximum number of simultaneous connections that can be set up by a matrix switch is equal to either the number of inlets or the number of outlets, whichever is the smaller.

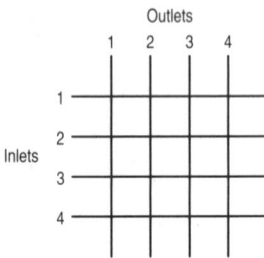

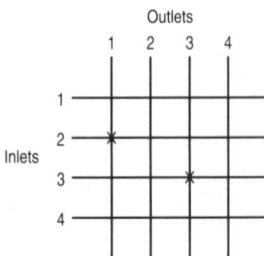

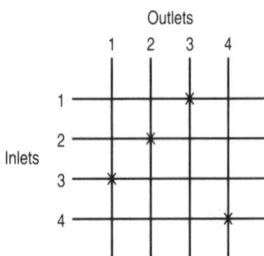

Fig. 10.13 (a) Switching matrix; (b) and (c) possible connections.

Example 10.1

A matrix switch has 15 outlets and can carry a maximum of 12 simultaneous calls. Calculate (a) the number of inlets, and (b) the number of crosspoints on the switch.

Solution
(a) Number of inlets = 12. (*Ans.*)
(b) Number of crosspoints = 12 × 15 = 180. (*Ans.*)

The efficiency of a matrix switch is reduced as the size of the switch is increased, and to increase efficiency switching is often arranged in two or more stages, as shown in Fig. 10.14.

It would be uneconomic for every telephone connected to a local exchange to be directly connected to a switch inlet. Since all the customers on an exchange will not attempt to make a call at the same time, and because most telephone calls are of fairly short duration, **traffic concentration** may be employed. This term means that a caller is only allocated a switch when he or she makes a call; there may be, for example, 10 switches provided per 100 telephone lines.

A digital telephone exchange

The basic block diagram of one kind of digital telephone exchange is shown in Fig. 10.15. The scanner continuously scans every line

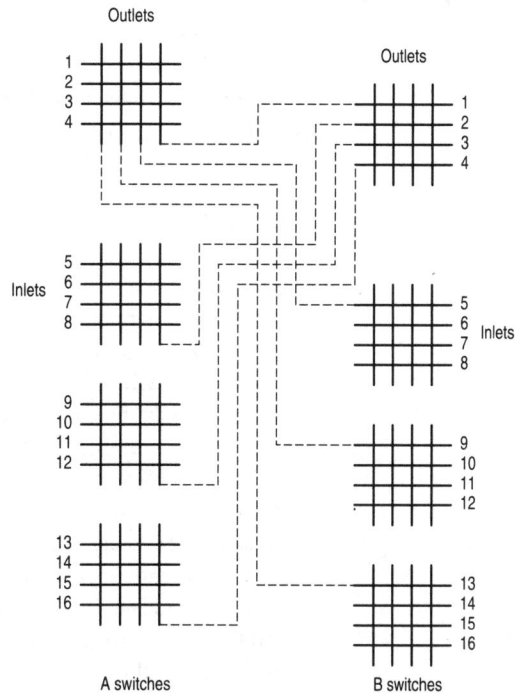

Fig. 10.14 Arrangement of matrix switches.

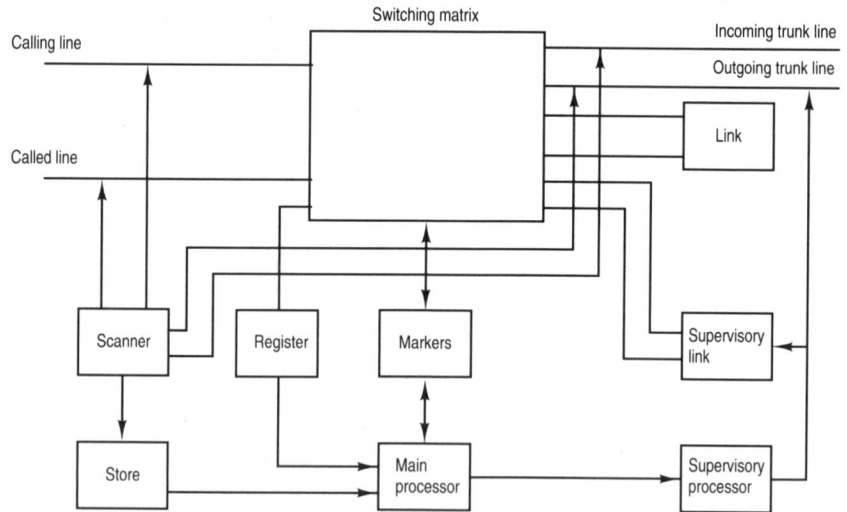

Fig. 10.15 Digital telephone exchange.

connected to the exchange to determine what type of line it is and whether its current status is free or busy. This information is passed to the store where it can be accessed at any time by the main processor. The scanning process is carried out at a very high speed so that up-to-date information is always available. When a local telephone user makes a call, the line is connected to a switch and the main processor is notified by the store that a particular line is being used. The main processor takes control of the setting up of the call until the point at which the called number is being rung; at this point control of the connection is passed to the supervisory processor. The main processor is able to handle many calls at the same time.

The main processor then finds a free register and informs it of the switch to which the calling line is connected. The markers are then instructed by the main processor to set up a connection between the register and the caller's switch. This connection is made via a link, as shown in Fig. 10.16(a). The first marker to set up a caller—link—register connection reports to the main processor and, unless this connection happens to be faulty, this link is used. The other markers can then turn their attention to the setting up of further connections within the switching matrix. Once the caller has been connected to a register, the dialling tone is returned and he or she may then 'dial' the wanted number. A register is needed to avoid the need to dial different codes to call the same telephone exchange from different locations within the telephone network. The digits representing the dialled number are passed into the register and, if it is a call to another telephone exchange, the exchange code part of the dialled number is converted into the numbers required to route the call through the trunk network to the destination exchange. The main processor then interrogates the store to determine whether the called local line, or an outgoing trunk, is free. If it is free the markers are instructed to

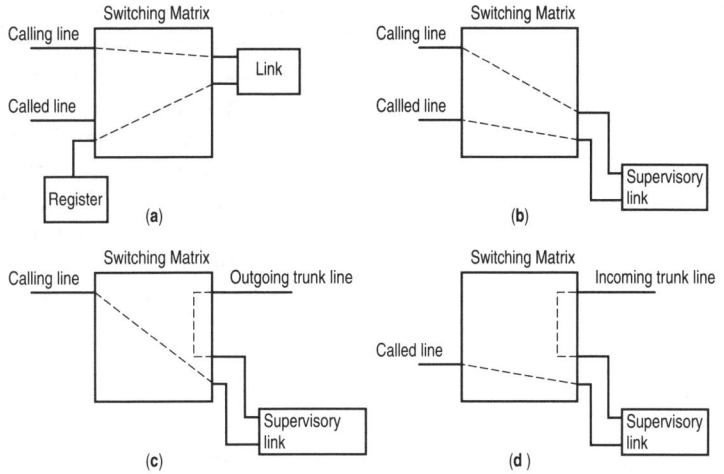

Fig. 10.16 Operation of a digital telephone exchange.

set up the required connection between the calling line and the called line but this time using a supervisory link. This step in the process is shown by Figs 10.16(b), (c) and (d) for a local call, an outgoing trunk call, and an incoming trunk call respectively. When the call has been made to a telephone connected to the same exchange. ringing current is applied to the called line. In all cases, once a connection has been established control of the connection is transferred from the main processor to the supervisory processor. This circuit performs such tasks as timing and charging for a call, and the release of the connection when the call is terminated.

Inter-exchange connections

The majority of telephone calls are made to a telephone that is connected to another telephone exchange. Telephone exchanges are connected to one another, either directly or indirectly, by lines known as **trunks**. The basic arrangement of an inter-exchange connection is shown in Fig. 10.17. When the caller dials the wanted number the control circuitry detects that the call is to a telephone connected to another telephone exchange. The controller then decides how the call

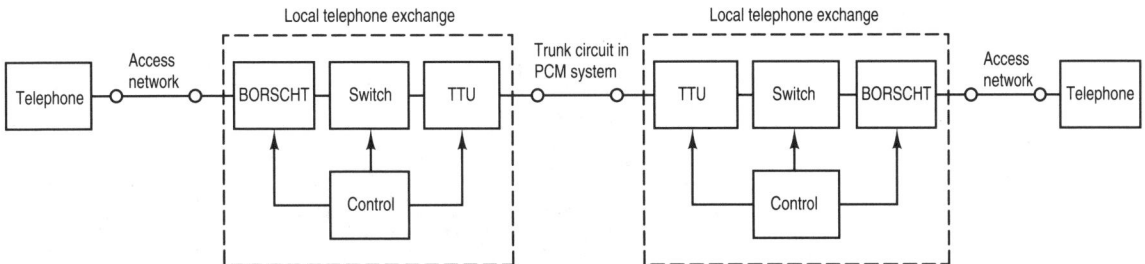

Fig. 10.17 Interconnection of two digital telephone exchanges.

should be routed through the network and selects an outgoing trunk circuit. The **trunk terminating unit** (TTU) provides trunk signalling, so that the remaining dialled digits can be sent to the other exchange, and interfaces to a PCM system that provides a channel to carry the trunk circuit. Once the link to the destination exchange has been established the incoming trunk at that exchange is switched to the wanted local line. The called number is then tested to see if it is free; if it is, then ringing current is applied to the line and ringing tone is returned to the caller. If the called line is engaged then busy tone is returned to the caller.

Routeing of calls between different telephone exchanges

It is not economically possible to directly interconnect every telephone exchange in a telephone network to every other exchange. A direct route is only provided when there is sufficient telephone traffic to justify the cost of provision. The remainder of the trunk traffic is routed via the **integrated digital network** (IDN).

The IDN is a telephone-switching network in which all the telephone exchanges are digital types and all trunk circuits are routed over PCM systems. The IDN consists of a large number of **digital local exchanges** (DLEs) and 53 **digital main switching units** (DMSUs). The basic layout of the IDN is shown in Fig. 10.18; each DLE has a direct link to its own DMSU and may also have direct connections to one or more DLEs. All the DMSUs are fully interconnected. When a telephone call is made from a telephone on one exchange to a telephone on another exchange the call may go via a direct link between the two exchanges. Nearly always the direct routes are

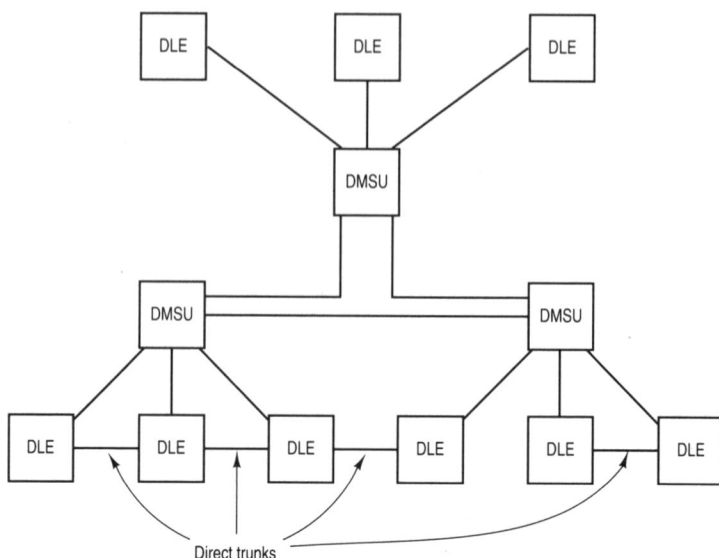

Fig. 10.18 The integrated digital network.

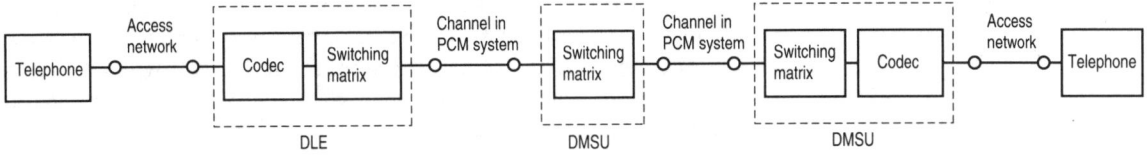

Fig. 10.19 A typical connection in the IDN.

between nearby exchanges because only then will there be enough telephone traffic to economically justify the provision of the direct route. In all other cases the call is first routed from the DLE to the DMSU; here it may be switched to another DLE or to another DMSU and from there to the destination DLE. In a DMSU the action is essentially to switch a channel in one PCM system to another channel in another PCM system.

The circuits between DLE, between DLE and DMSU, and between two DMSUs are routed over multi-channel PCM systems. The capacity of the basic PCM system is 30 telephone channels and the multiplexing process — i.e. grouping several telephone channels to work over a single transmission path — is carried out as part of the digital exchange switching action. The set-up of a typical call is shown in Fig. 10.19. Some of the possible connections that might be set up in the telephone network are shown in Fig. 10.20.

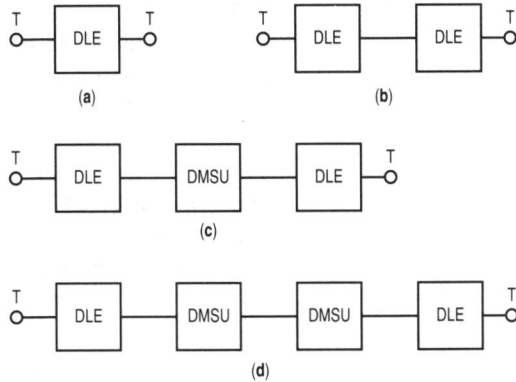

Fig. 10.20 Possible IDN connections.

Office telephone systems

A number of extension telephones can be connected to a 'call-connect' or telephone system that provides both for internal calling between extensions and for calls to and from the PSTN. A wide variety of office telephone systems are readily available, ranging in size from 1 exchange line and 4 extensions up 32 exchange lines and 144 extensions. In addition, specialized installations can be designed and installed for particularly large offices. Each extension is able to 'dial' every other extension and to make and receive outside calls. To obtain an outside line a preliminary number, often 9, must first be 'dialled'. If required, some extensions can be barred from making outside calls.

Although older office telephone systems were often manually operated all modern systems are automatic; the **private branch automatic exchange** (PABX) or **central control unit** (CCU) will vary in its physical dimensions from a small wall-mounted box to a cupboard-sized floor-standing cabinet.

All systems may be set up 'switchboard-style' with a dedicated operator or arranged so that any extension is able to answer calls from the PSTN. An operator's console has a number of keys each of which provides some kind of service, and a standard 0 to 9, plus * and #, keypad. One or more LCD display panels are provided which may give instructions and/or guidance, show the current status of each extension, i.e. free or engaged, and enable the progress of a call to be tracked.

A number of different facilities may be offered by an office telephone system and the more common of these facilities are listed below.

1. **Group pick-up**. All the extensions are assigned to a particular group. Depending upon the size of the installation there may be one, two, or even more groups. If an extension in a group is aware that another extension in that group is being rung but is not answering, that extension can answer the call by 'dialling' the group pick-up code. This code is typically a two-digit number such as 66. In another version of this facility all the telephones in a group will ring simultaneously when there is an incoming call and any of the extensions may answer.

2. **Trunk wait-on-busy**. This facility allows any extension to extend an external call to a busy extension. The busy extension is informed that a call is waiting by hearing two short bursts of a tone. If, after a short time, the busy extension does not take the call, the call is automatically routed back to the original extension.

3. **Extension wait-on-busy**. This facility allows a busy extension to be alerted that another call is waiting for it. The busy extension is alerted by hearing one burst of tone.

4. **Call diversion**. Calls can be diverted from one extension to another by dialling the 'divert all calls' code followed by the number of the extension to which the calls are to be diverted. It is often possible for the first extension to still be able to make outgoing calls but no incoming calls will be received.

5. **Diversion on busy**. If the 'diversion on busy' code followed by another extension number is dialled any incoming calls that arrive when the first extension is busy will be diverted to the other extension.

6. **Three-party connection**. This facility allows an extension to place a call on hold and to originate another call to another extension. The new extension can be spoken to privately and the original call can then be continued or transferred to the other extension.

7. **Call back on busy**. If an extension calls another extension that turns out to be busy, the 'call back on busy' feature can be activated. When the called extension becomes free it is automatically rung and the calling extension gets a distinctive ring so that it knows that its earlier call is now being completed.
8. **Hold**. The hold facility allows an extension to hold a call and, while it is held, make another call.
9. **Night service**. This facility allows all incoming calls after office hours to be diverted to a particular extension.
10. **Messaging and paging**. The messaging facility alerts an extension to the fact that a message has been left asking the caller to call back later, and the paging facility allows a person to be paged via telephone loudspeakers.

Other facilities that are provided with most systems include:

- Abbreviated dialling — regularly dialled numbers can be given an abbreviated code to give faster dialling.
- Account codes — a code can be allocated to each customer and then any calls made on their behalf are automatically timed and recorded for later billing.
- Call barring — some, or all, extensions can be barred from making external calls.
- Conference calls — a three-way conversation can be established between three extensions or between two extensions and one outside line.
- Different ringing tones for internal and external calls.

Call sequencer

To avoid the possible loss of business arising from not answering all incoming calls quickly enough, a **call sequencer** can be employed. This equipment automatically answers and welcomes every caller and holds them in sequence until their call can be answered. Music can be played to ensure the caller that, although he or she is waiting, the call is still connected.

Bounded transmission media

Signals are transmitted from the telephone to the local telephone exchange and between telephone exchanges over some kind of bounded transmission medium. The media that are used include open-wire pairs, audio copper conductor cable, coaxial and optical fibre cable, and microwave radio systems — and all have their relative advantages and disadvantages.

The telephones in people's homes and in small offices are connected to a nearby distribution point mounted on top of a telephone pole by an open-wire pair. This connection method has the advantage of being very cheap, but it is really only feasible for one or two pairs of conductors. Its transmission performance is unpredictable and variable largely because the spacing between the two conductors is always changing as the wires swing in the wind and are subject to moisture. The connections between the distribution points and the local telephone exchange are made by multi-pair audio cables. Each pair consists of two copper conductors that are separated from one another, and from other pairs, by a plastic insulating material such as polythene. A large number of pairs may be combined using either an arrangement known as 'star quad' or another arrangement known as 'unit twin'. Audio-frequency cables have a limited frequency response and are subject to crosstalk, which means that they are unsuited for the carrying of multi-channel PCM systems. Audio cables are now used only in the access network.

The trunk network in the UK is routed over a combination of coaxial and optical fibre cable and microwave radio systems. Optical fibre cable now provides much the greater total mileage. Both coaxial and optical fibre cables are able to provide a very wide bandwidth which allows them to transmit PCM systems. Optical fibre, which is a much more recent introduction than coaxial cable, has a number of advantages that have led to its rapid introduction into the trunk network. The most important of these advantages are that it has a very wide bandwidth, it is immune to electromagnetic interference, it is of low loss, and it is relatively cheap.

11 Data communication

Frequently there is a need for data to be transferred from one mainframe computer to another, or between a mainframe computer and a data terminal (which may itself be a PC), or between two PCs, or between a computer (mainframe, mini, or PC) and a peripheral. Very often the two locations between which the data is to be transferred are not at the same site and, indeed, may even be in different countries. A data communication link that is able to transmit data without error between the two locations is then necessary. A data circuit may be set up when required using the **public switched telephone network** (PSTN), or the **public switched data network** (PSDN), or a private switched data network. Alternatively, a permanent **dedicated** data link can be established between two points using circuits allocated by the telephone administration.

A private data network that covers a wide geographical area is known as **wide area network** (WAN). The choice between using public or private circuits or a combination of both, and whether the WAN should be dedicated or switched or perhaps a mixture of both, is made after due consideration of such factors as costs, availability, speed of transmission, and overall performance. In some cases immediate access to a computer is of prime importance. Take, for example, the well-known bank or building society cash dispensers; each of these machines requires a dedicated line to the bank's or the building society's computer so that customer's demands for cash may be rapidly satisfied. This is an application where the inherent delays associated with setting up a connection via the PSTN would not be acceptable.

A data network that serves a number of computers, peripherals, and terminals located within a small area, such as a single building or a factory site, is known as a **local area network** (LAN).

Inside a computer data is moved around in groups of 8, 16, 32 bits, or more, depending upon the type of computer. For a very short connection to a peripheral it is possible to transfer data using **parallel transmission** in which all the bits forming a character are simultaneously transmitted. This means that a multi-conductor cable is used as an n-bit parallel interface to connect the computer to the peripheral, as shown in Fig. 11.1. Parallel transmission of data is very fast but the cable employed is very expensive, and as there are also some technical difficulties **serial transmission** of data is used

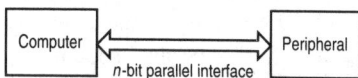

Fig. 11.1 Parallel transmission of data.

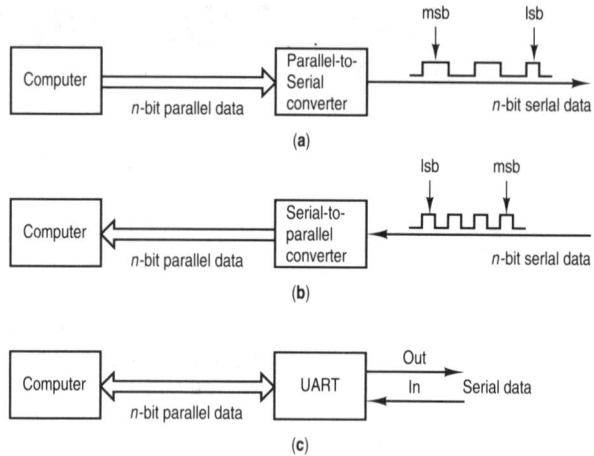

Fig. 11.2 Serial transmission of data: (a) output from computer; (b) input to computer; and (c) use of a UART.

for all but the shortest connections. The basic idea of serial transmission is shown in Fig. 11.2. In Fig. 11.2(a) the output data of the computer, which is in parallel form, is applied to a **parallel-to-serial converter** and this circuit converts the data into serial form. Serial form means that the bits making up a character are transmitted in sequence, with the least significant bit (lsb) transmitted first and the most significant bit (msb) transmitted last. The **data transfer rate** is considerably reduced. In the opposite direction of transmission, see Fig. 11.2(b), the incoming serial data is applied to a **serial-to-parallel** converter and here the data is converted into parallel form. The two converters are always provided within the same interface IC which is given a variety of different names by various manufacturers (p. 144). One of these names is UART, and this name has been used in Fig. 11.2(c). The interface IC is normally mounted inside the computer and its serial input/output terminals are connected to the computer's serial port. The serial transmission of data uses one of a number of standard formats that have been specified by the Electrical Industry Association (EIA). The most common of these formats is the EIA 232 specification which has now reached version E.

To ensure that the received data is sampled by the receiver at the correct instants in time, serial data must have some synchronization bits added to it, plus, usually, some further bits to provide detection of any error that might occur.

The transmission of data over the PSTN

Because the provision of permanently leased, or dedicated, circuits to carry data is not viable unless the traffic is sufficiently high to economically justify the provision and the delay inherent in setting up a connection is tolerable, many data circuits are established using the PSTN. The PSTN was originally designed for the transmission of analogue speech signals and although the trunk network is now entirely digital in its operation the access network is still analogue.

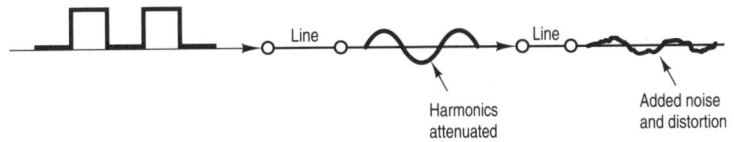

Fig. 11.3 Data signal distorted by a transmission line.

A digital data signal cannot be directly transmitted over the PSTN because:

- The local line in the access network may include a transformer to match audio-frequency local cables together and these will not pass the d.c. component of the data signal.
- Attenuation causes the waveshape of the data signal to become distorted, as shown in Fig. 11.3; this is an effect that increases with both the length of the local lines and the bit rate.
- The bandwidth of the PSTN cannot transmit the fundamental frequency and lower-order harmonics of low-bit rate signals, or the harmonics of high bit-rate signals. In both cases further distortion of the data signal will occur. When a received data signal is distorted there is an increased probability that the data will be mis-read by the receiver and an error produced.
- The d.c. component of a data signal cannot be transmitted by either digital telephone exchanges or the PCM trunk network.

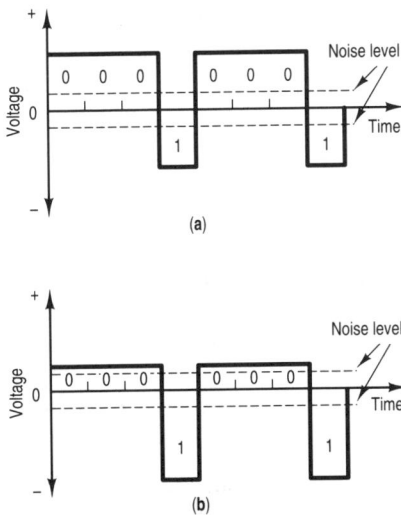

The effect on a digital data signal of removing its d.c. component is shown in Figs 11.4(a) and (b). The d.c. component is equal to the average value of the data waveform; if this component is removed the average value will become zero volts and the whole waveform will move in the negative direction. It can be seen from the figure that this movement increases the voltage of the 1 bits but reduces the voltage of the 0 bits until they are not much bigger than the noise level. The receiver is then likely to have difficulty in correctly recognizing each of the received bits and the likelihood of error will increase.

Fig. 11.4 (a) Data signal; (b) the same signal with its d.c. component removed.

Modems

If the PSTN is to be used for the transmission of data signals, those signals must first be changed from digital form into the equivalent analogue form. At the far end of the connection the received analogue signal must be converted back into its original digital form so that the computer, or data terminal, will be able to understand it. The necessary digital-to-analogue and analogue-to-digital conversions are carried out by a **modem**. The basic circuit of a PSTN connection using two modems is shown in Fig. 11.5. The serial digital data output of the UART in the computer is applied to the modem and here it modulates an audio-frequency carrier to produce a digitally modulated signal which is in the bandwidth provided by the PSTN (300 to 3400 Hz). The voice-frequency (VF) signal is then transmitted over

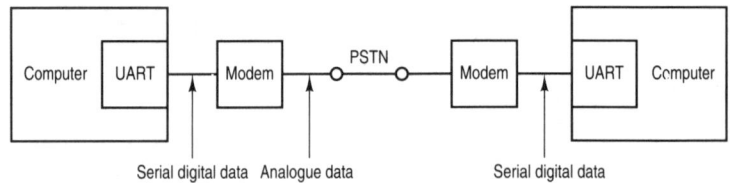

Fig. 11.5 Use of modems.

the PSTN, in exactly the same way as a speech signal, to its destination. Here another modem demodulates the received VF signal to convert it back into its original digital form. The digital data signal is then applied to the data terminal which may be another computer (as shown), a visual display unit (VDU), or a database.

A number of other facilities may also be provided by a modem.

1. **Auto-'dialling'** and **auto-answering**. Auto-'dialling' means that the modem has the ability to handle all the actions that are necessary to set up a telephone call over the PSTN. The modem is able to recognize dialling tone, engaged tone and ringing tone and knows when it should start sending data and when it should clear down a call. If a called number is busy the modem is able to remember the 'dialled' number and continue re-'dialling' it until the wanted connection has been set up. Auto-answering means that the modem is able to answer an incoming call without any help from an operator and set itself up to process the received data.

2. **Error detection and correction**. Some modems are provided with the ability to detect and correct errors in received data.

A modem is mainly described by the maximum bit rate at which it is able to operate. The various bit rates that are used have all been standardized by the CCITT V series of recommendations which also specify the type of digital modulation to be employed. The manufacturers of modems generally quote the appropriate V numbers, and these are listed in Table 11.1.

V21 was the first dial-up standard to be introduced but it is too slow for most modern applications. V22 is another early standard but it

Table 11.1

Recommendation	Bit rate (bits/s)	Modulation method	Recommendation	Bit rate (bits/s)	Modulation method
V21	300	FSK	V27	4 800	DPSK
V22	1 200	DPSK	V27 bis	4 800	DPSK
V22 bis	2 400	QAM	V29	9 600	QAM
V23	1 200/75	FSK	V32	9 600	QAM
V26	2400	DPSK	V32 bis	14 400	QAM
V26 bis	2 400	DPSK	V33	14 400	QAM
V26 ter	2400	QAM			

has now mainly been replaced by either V22 bis or V32. V22 bis is often used since it gives a reasonably high speed at relatively low cost. V32 is much faster than V22 bis and uses more expensive modems; it is also often used for access to databases where there may well be a considerable amount of data to be transferred. V32 bis and V33 are even faster and more expensive standards. V23 offers different speeds in the two directions of transmission and is intended for use with such systems as PRESTEL. High-speed modems are often used in conjunction with error correction and data compression schemes, V42 and V42 bis respectively, to get up to a four-times increase in speed. The apparent maximum bit rate is then $4 \times 14\,400$ or $57\,600$ bits/s but the serial output ports on PCs have a maximum bit rate of $38\,400$ bits/s.

Many modems are able to work at more than one bit rate. The calling modem starts at the highest speed it can manage and determines whether the other modem is able to cope. If it cannot, then the bit rate is reduced until the maximum speed at which both modems are able to operate is discovered. Data transfer can then commence.

For modems to be able to perform their various tasks they must be under the control of a computer and this means that the computer must be supplied with communications software. A communication program formats and checks data before and after transmission and allows all communications to be controlled from the keyboard. Most computers use the **Hayes AT** command set.

The transmission of data over the PSDN

The public switched data network (PSDN) consists of a number of **packet-switching exchanges** (PSE) that are interconnected by high-speed data links. A user of the PSDN is connected to the nearest PSE by a **dataline**. Often a dataline is a leased dedicated analogue telephone line having a modem at each end, or it may be a dial-up connection via the PSTN. The British Telecom version of PSDN is known as Packet Switch Stream (PSS) within the UK, and as International Packet Switch Stream (IPSS) for overseas circuits.

The use of the PSDN gives a user two advantages over the use of the PSTN. These are (a) lower call charges and (b) a higher performance because the PSDN is able to provide error protection from one end of the network to the other. Private packet switching systems are also employed by various large organizations.

Wide area networks

A **wide area network** (WAN) is a private network of computers and data terminals that covers a large geographical area. If the data traffic between two locations economically justifies the cost of provision, the best way of interconnecting two computers, or a computer and a data terminal, that are located at different sites is by means of a dedicated circuit. A leased dedicated circuit can have its transmission performance optimized for the transmission of data and may, perhaps, be an all-digital circuit. For all required links where the data traffic is smaller connection to a switched private network may be the best solution. For occasional communications, however, it is much cheaper to use a dial-up connection via either the PSTN or the PSDN.

Private data networks are used in the UK by many large organizations such as the clearing banks, the building societies, the gas, electricity and water companies, the railways, Government departments, and insurance companies. Many of the networks use lines leased from British Telecom, but some use lines leased from Mercury, and others, for example, use lines installed alongside railway tracks. Most of the lines employed are now digital but the access to them is very often via a local analogue line. This means that a modem is required at either end of the line. If the local lines are conditioned to provide a high-speed circuit it is possible to transmit several data circuits over a single line using multiplexing. Figure 11.6 shows a data circuit that provides four circuits from a computer to four distant data terminals. The multiplexing is carried out on a time division basis. The computer uses a polling technique in which it addresses each data terminal in turn to see if that terminal has any data to exchange; if it has, data is then transmitted. When a data exchange with one terminal has been completed the computer moves on to the next data terminal and interrogates it to see whether this terminal has data to exchange, and so on. The polling action is carried out very rapidly.

Another way in which a line can be made to carry data from several circuits is shown in Fig. 11.7. A **cluster controller** is used to split

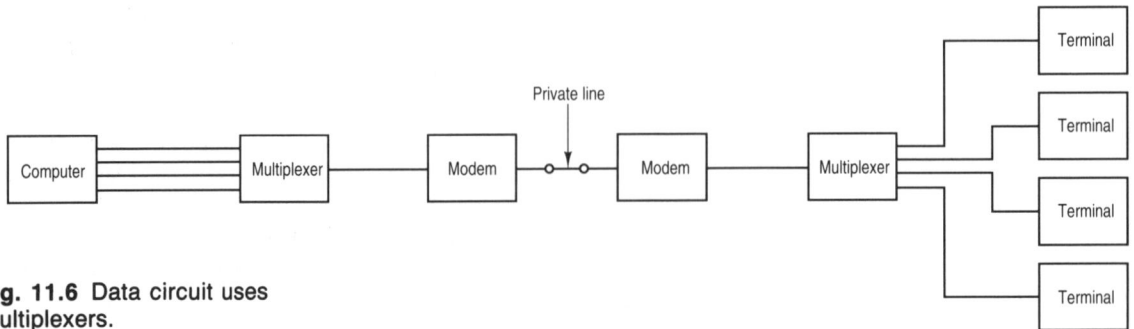

Fig. 11.6 Data circuit uses multiplexers.

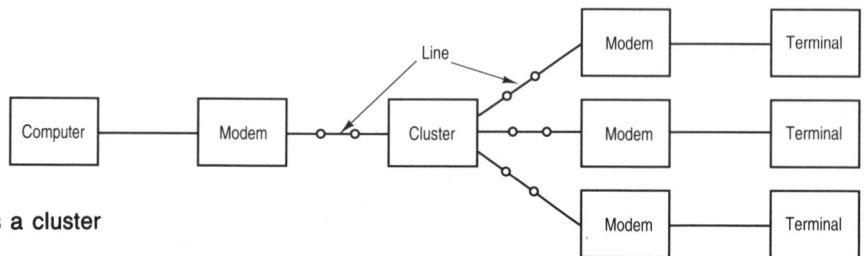

Fig. 11.7 Data circuit uses a cluster controller.

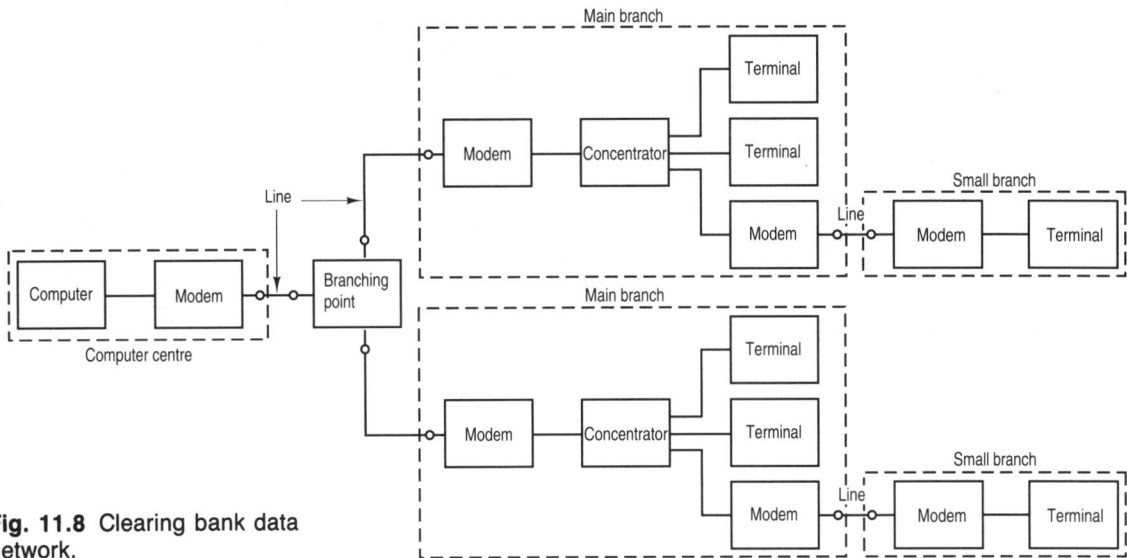

Fig. 11.8 Clearing bank data network.

the main line path into two or more paths. The data terminals or PCs that are connected to the cluster controller must be located within a few kilometres of the junction. Again the mainframe computer polls each data terminal in turn to discover whether or not the terminal has data to be transferred.

Figure 11.8 gives the block diagram of a typical data network for a clearing bank. The network gives both major and minor branches of the bank access to the bank's central computer. Branching points are used to split the main high-speed line between different main bank branches; this technique saves on line costs. At a major branch the incoming data circuit is connected to a **concentrator**. A concentrator acts in a similar way to a multiplexer and allows several low-speed data circuits to operate over a single higher speed circuit. Some of the low-speed data circuits are connected to data terminals at the major bank (one or more may be a cash dispenser) and some are connected to another modem for data transmission to a minor branch of the bank. The data circuits connecting the major and minor bank branches work at a lower bit rate than the main data circuit.

Use of digital circuits

British Telecom offers high-speed data circuits, known as Kilostream and Megastream, that can be used in a private data network. A Kilostream circuit is extended into the customer's premises and it terminates there on a **network terminating unit** (NTU). The function of a NTU is to convert the PCM signals used on the Kilostream circuit into the standard EIA 232 signals used by the computer's serial

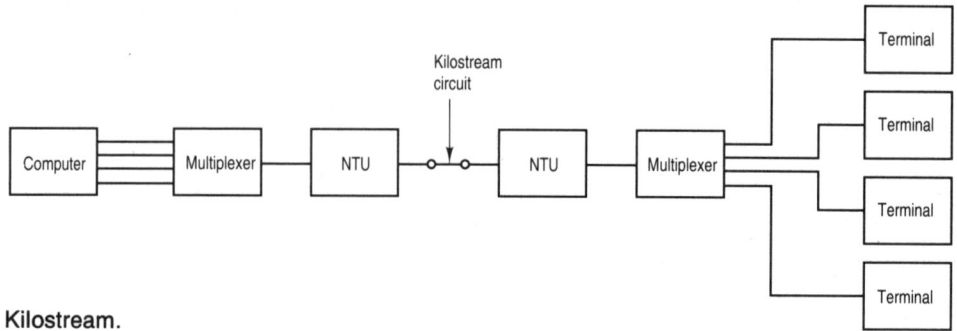

Fig. 11.9 Use of Kilostream.

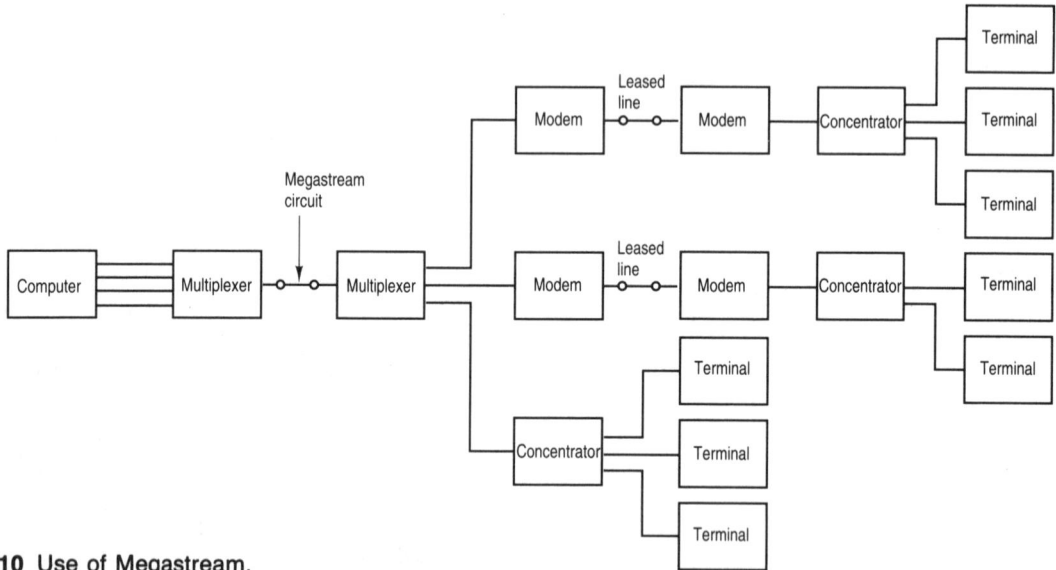

Fig. 11.10 Use of Megastream.

interface. The use of Kilostream in a typical data network is shown in Fig. 11.9. Kilostream operates at 64 kbits/s and provides either one 64 kbits/s data circuit or, using multiplexing, a number of 2400, 4800, or 9600 bits/s data circuits. Megastream operates at either 2 Mbits/s or 8 Mbits/s and provides a high-speed circuit that can be used for data signals or for both speech and data signals. An example of the use of a Megastream circuit is given in Fig. 11.10.

Many modern WANs use packet-switching technology either entirely or mixed in with the methods outlined.

Local area networks

A **local area network** (LAN) is a data network that operates over a small geographical area such as within a single building or on a factory site. The building, or site, must be wired up for the LAN using either twisted pair cable, coaxial cable, or fibre optic cable. A LAN is used to interconnect a number of PCs and to allow them access to centralized files and to expensive facilities such as laser printers and plotters whose provision cannot be justified economically for a single user. A LAN may also give PCs access to one or more mainframe computers and/or minicomputers and/or to other LANs.

Most LANs are of one of two topologies: Ethernet or 'token ring'. Usually, a **network server** is required that acts to manage the operation of the entire network. The network server is a powerful PC with large memory and a high-capacity hard disk; it is provided with network software to control the operation of the LAN. The other types of server, also PCs, that may be employed are a **file server** which acts as a central disk storage device, a **database server** whose function is to run a large database, a **communications server** which provides access to other networks and/or to a mainframe computer, and a **printer server** which is a PC that is provided with software that allows other PCs to use the printer connected to it just as though the printer were connected to those other machines.

One example of the use of a LAN occurs in a typing pool; a number of VDUs are given access to files of standard letters, names and addresses of regular customers, customer's account details, and so on. When letters have been prepared they can be printed out by a common printer to which all VDUs have access via a printer server.

The simplest form of LAN is what is known as the **peer-to-peer** network. It is so-called because every PC in the network is able to access every other PC and any PC can be configured to allow its resources to be used by any other PC. All the PCs must be provided with network software. Usually at least one PC is designated to be a file server and another PC is designated to act as a printer server. The file server maintains a database of all the information likely to be required by the various PCs and it also runs the network. The server PCs are installed with the appropriate software to allow other PCs to use their hard disk, or their printer. If a particular PC has its own hard disk, labelled C as always, then the hard disk accessed via a file server would be labelled D (and E if there were two file servers). When a file is printed out the server's software saves all the data and informs the originating PC that the data has been printed. There is no risk of two lots of data being printed at the same time and getting mixed up because the server software queues the printing requests and prints them in sequence — which is known as **print spooling**.

A peer-to-peer LAN is resilient and cheaper than the alternative method of operation; its main application is as a file and printing service for a small group of PCs. Most peer-to-peer LANs use a technique known as **Ethernet** and are able to transfer data at bit rates up to about 1.25 Mbits/s. Figure 11.11 shows the layout of a possible peer-to-peer LAN.

The alternative method of operating a LAN is a server-based network that uses a dedicated **network operating system** (NOS). The NOS requires a dedicated PC to act as the system server and this PC must be fitted with a large, fast-access, hard disk which is used to

```
┌──────┐   ┌──────┐   ┌──────┐   ┌──────┐
│  PC  │   │ File │   │  PC  │   │Printer│
│      │   │server│   │      │   │server│
└──┬───┘   └──┬───┘   └──┬───┘   └──┬───┘
───●─────────●─────────────●────────●──────────●──────
          ┌──┴───┐           ┌──┴────┐   ┌──┴───┐
          │  PC  │           │Network│   │  PC  │
          │      │           │server │   │      │
          └──────┘           └───────┘   └──────┘
```

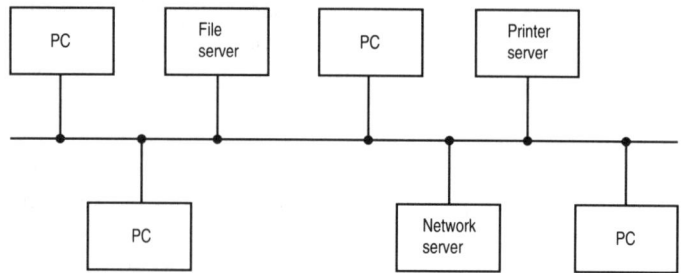

Fig. 11.11 Peer-to-peer LAN.

store data and programs that any PC in the network can use. The PC should be powerful enough to handle the expected workload and should have at least 4 Mbytes of RAM. Other PCs on the LAN are able to run programs that are stored on the server as well as using the data stored at the server. On receipt of a request from another machine the system server will read a program held on its hard disk and then send this program over the network to the requesting PC where it can then be run. It is possible for several different PCs to be running different programs obtained from the server at the same time without any interference. Sometimes some, or all, of the other PCs do not have hard disks of their own which makes them much cheaper to buy. A NOS is able to support a large number of PCs, frequently up to about 200. A NOS LAN may also use the Ethernet technique or, instead, another technique known as **token ring**; both methods can work at bit rates of up to about 10 Mbits/s. In a NOS LAN the individual PCs are all 'clients' of one, or more, central servers which provide the services for the whole network. This kind of LAN is more expensive than the peer-to-peer type and the dedicated file and printer servers can be optimized for their tasks. The disadvantages of a NOS LAN are that if there is only one server, and it becomes faulty, the whole network will fail and the individual PCs will be unable to access the resources of other individual PCs.

The arrangement of a typical token ring network LAN is shown in Fig. 11.12. A hub, or wiring concentrator, is provided with either four or eight ports that are used to connect PCs, plus ring-in and ring-out ports that are used to connect with any other hubs in the network. Only one hub is required if the number of PCs to be interconnected is not greater than eight and no PC must be located at a distance in excess of about 300 m from the hub. If more than one hub is required but all the PCs connected to them are quite near, then all the hubs may be located at a single location. The hubs can then be placed on top of each other as shown in Fig. 11.13.

There are two other technologies in use for networking; namely, Apple's LocalTalk and Fibre Distributed Data Interface (FDDI), the latter having been designed for use with fibre optic cables. FDDI can operate at bit rates as high as 100 Mbits/s.

A LAN is often employed in a supermarket or a department store where every electronic till acts like a PC to control and monitor the

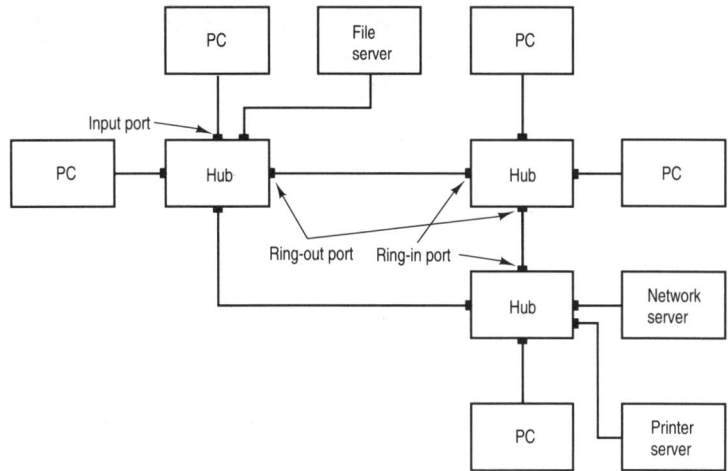

Fig. 11.12 Token-ring LAN.

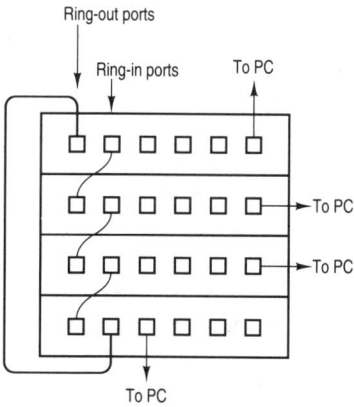

Fig. 11.13 Hubs at a single location.

daily sales. Each time an item is sold its details are entered into the till, by hand or by scanner, and the details are sent to an in-store large PC, or even a minicomputer, which in turn is a node on a WAN. The large PC, or minicomputer, reports back to a central database whose associated mainframe computer polls all the stores connected to the system nightly. The database is continually up-dated as new data comes in and it can be linked to a spreadsheet to allow calculations to be carried out to keep track of stock levels. The system can easily be arranged so that further supplies of goods can be automatically ordered as stock levels fall below a predetermined figure. This is how supermarkets are able to arrange daily deliveries of fresh food in the correct quantities. Most of the food that is delivered goes straight onto the shelves and is re-ordered on a daily basis so that each day fresh food is available to shoppers. As each shop has permanent access to the central computer, the price of goods can be altered as often as required.

12 Radio systems

A radio system is used to transmit information from one place to another by means of electromagnetic radio waves. The information may be transmitted from a central point to a large number of receivers located in all directions from the transmitter to provide a **sound broadcast** radio system. In the United Kingdom sound broadcasting stations are operated by both the BBC and various commercial organizations. BBC programmes are transmitted in the long and medium wavebands and in the VHF band, both nationally and with just local coverage. Most commercial radio stations provide a local radio service in the medium and VHF bands but there is also one national commercial station, also in the VHF band. Sound broadcasting provides a one-way service, i.e. there is no provision for two-way conversation. Other kinds of radio systems provide two-way communication between two fixed locations; most systems of this kind carry a multi-channel telephony system with perhaps hundreds or even thousands of circuits. Such systems are employed as an integral part of the national and international telephony networks. A third kind of radio system is mobile radio. A mobile radio system provides communication between one or more central control stations and a number of mobile radio-telephones. Mobile systems may be land, maritime, or aero, and land systems may be either private or linked to the public telephone service.

In all kinds of radio system a baseband signal modulates a radio-frequency carrier wave to position the modulated signal at the allotted part of the radio-frequency spectrum before the signal is radiated into the atmosphere. The radiated radio wave propagates through the atmosphere to its destination and here it is intercepted by a receiving aerial. The signal picked up by the aerial is passed onto a radio receiver and the receiver recovers the information from the modulated wave and applies it to a loudspeaker, or to a telephone line, or to some other receiving device.

Radiation from an aerial

When a current flows in a conductor a magnetic field is set up around that conductor. The direction of the magnetic field is determined by the direction in which the current is flowing, and the intensity of the field is directly proportional to the amplitude of the current. If the direction in which the current is flowing is reversed, the magnetic

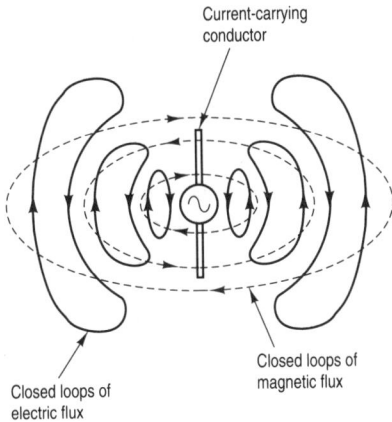

Fig. 12.1 Radiation of electromagnetic wave from an aerial.

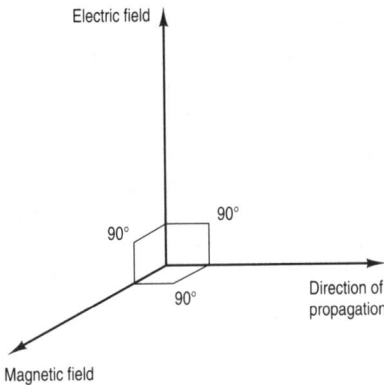

Fig. 12.2 Electric and magnetic field directions.

field will first collapse into the conductor and then build up in the opposite direction. As the magnetic field is changing — decreasing or increasing — it generates a changing electric field. In turn, the changing electric field generates a changing magnetic field.

If an alternating current flows in a conductor it will continually flow first in one direction and then in the other. The magnetic field set up around the conductor will then be continuously building up in one direction, collapsing, and building up in the opposite direction. The a.c. current-carrying conductor is thus surrounded by two continually changing fields, one magnetic and the other electric. The variations in the electric and magnetic fields do not take place instantaneously and, at frequencies higher than about 15 kHz, not all of the energy contained in a magnetic field has had time to return to the conductor before the next field is starting to build up in the opposite direction. Once this situation exists, the radio-frequency energy that has not managed to return to the conductor before the magnetic field changes its direction is now propagated away from the conductor with the same velocity as that of light, i.e. 3×10^8 m/s, and this is shown in Fig. 12.1.

As the frequency of the current flowing in the conductor increases, the time available for the magnetic field to collapse into the conductor becomes even smaller and so even more energy is radiated. This means that the amount of energy radiated from the conductor increases with increase in frequency.

The radiated energy is known as an **electromagnetic wave** and the conductor is known as an **aerial**. In an electromagnetic wave the directions of the magnetic and electric fields are always at right angles to one another, and also at right angles to the direction of propagation. This important point is illustrated in Fig. 12.2. The plane which contains the electric field is known as the **polarization** of the radio wave. The magnitude of the radiated radio wave is directly proportional to the frequency of the wave and is inversely proportional to the distance from the aerial. The radio wave propagates through the atmosphere and is able to pass through such insulating materials as brick, wood, and glass, and through a vacuum, but it is reflected by any conducting surface it might meet.

The electromagnetic spectrum is shown in Fig. 12.3, in which frequencies and wavelengths are given. It can be seen that radio

Fig. 12.3 The electromagnetic spectrum.

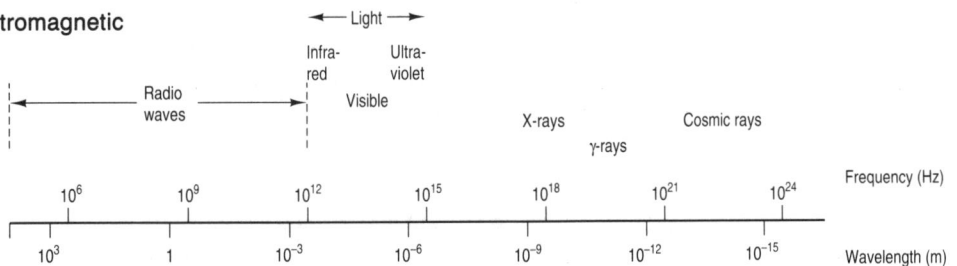

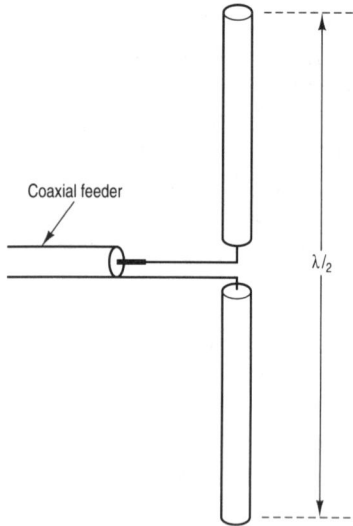

Fig. 12.4 The λ/2 dipole aerial.

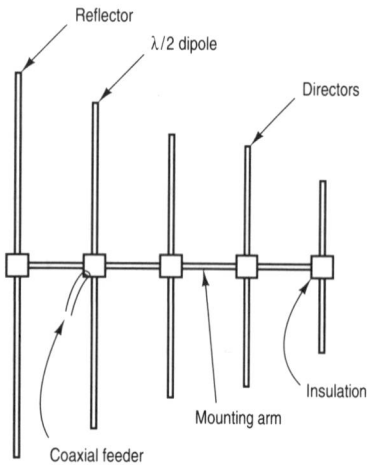

Fig. 12.5 Television receiving aerial.

Propagation of radio waves

Table 12.1

Frequency Band	Frequencies
Very low frequency (VLF)	Up to 30 kHz
Low frequency (LF)	30 to 300 kHz
Medium frequency (MF)	300 kHz to 3 MHz
High frequency (HF)	3 to 30 MHz
Very high frequency (VHF)	30 to 300 MHz
Ultra high frequency (UHF)	300 MHz to 3 GHz
Super high frequency (SHF)	3 to 30 GHz

frequencies occupy less than one-half of the spectrum: the higher frequencies or the shorter wavelengths are, in increasing frequency, infra-red, visible and ultra-violet light, X-rays, γ-rays, and cosmic rays. The radio-frequency part of the electromagnetic spectrum is divided into a number of **bands**, which are listed in Table 12.1.

The half-wave dipole

The simplest form of aerial is the **half-wave dipole** shown in Fig. 12.4. It consists of two lengths of cylindrical conductor whose total length is equal to one-half of a wavelength at the frequency of the signal being radiated. The two conductors are supplied with a radio-frequency current by a coaxial feeder; the outer conductor of the coaxial pair is connected to one half of the dipole while the inner conductor is connected to the other half. The λ/2 dipole can also be used as a receiving aerial. The λ/2 dipole aerial is used in many radio systems, sometimes on its own but more often in combination with one or more λ/2 dipoles or a number of other elements. The well-known television aerial, seen on many rooftops, is shown in Fig. 12.5; it consists of a half-wave dipole plus a **reflector**, and a number of **directors** (three in the figure).

The way in which a radio wave propagates through the atmosphere depends upon the frequency of the wave, because of the varying attenuation provided by the Earth's surface, and the varying characteristics of a part of the atmosphere known as the **ionosphere**. There are four main propagation paths over which a radio wave may travel. These are:

● Ground or surface wave
● Sky wave
● Space wave
● Communication satellite.

Ground or surface wave

At frequencies in the VLF and LF bands transmitting aerials must be physically very large and are mounted vertically upon the ground. Such an aerial transmits a radio wave that propagates very near to the surface of the Earth. The **ground wave** follows the curvature of the Earth as it travels and it is subjected to losses because of energy dissipated in the Earth. These losses increase rapidly with increase in frequency. Because of the frequency dependence of these ground losses, the ground wave has a useful range only at frequencies in the VLF, LF, and MF bands. Figure 12.6 shows a ground wave providing radio communication between two points well separated on the Earth's surface.

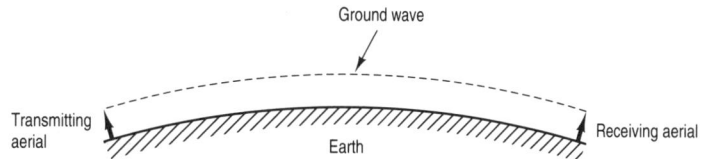

Fig. 12.6 Ground wave propagation.

Sky wave

The Earth is surrounded by a wide region of ionized gases known as the **ionosphere**. The height of the ionosphere is not fixed but usually it varies between limits of about 50 to 400 km. At frequencies in the HF band the ionosphere acts as a refracting medium to any radio waves that should enter it. An HF radio wave from the Earth's surface directed into the ionosphere is therefore progressively refracted as

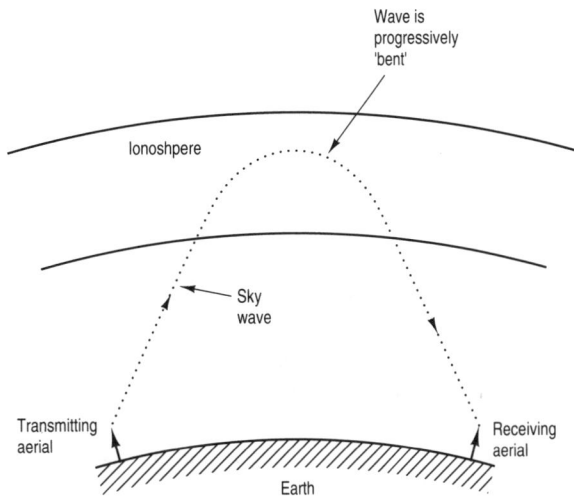

Fig. 12.7 Sky wave propagation.

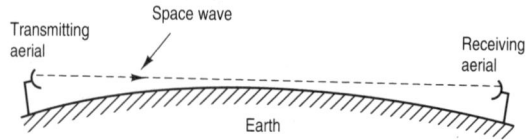

Fig. 12.8 Space wave propagation.

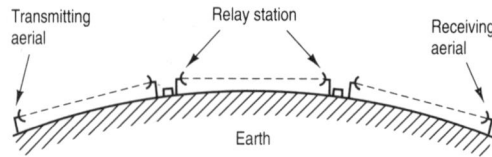

Fig. 12.9 Radio relay system.

it travels through the ionosphere and, if certain conditions are met, it may be returned to Earth. Sky wave propagation of a radio wave is shown in Fig. 12.7. A transmitting aerial is employed that is able to direct most of its radiated energy up into the sky and, if the angle at which the wave enters the ionosphere is correct, the wave will be returned to Earth at the location of the destination radio station.

Space wave

At frequencies in the VHF, UHF, and SHF bands the ionosphere has little effect upon a radio wave and so it cannot be used to return a wave to Earth. Instead, radio waves are transmitted over a straight 'line-of-sight' path between two aerials that are mounted on top of a mast or a tower. Space wave propagation is shown in Fig. 12.8. Since propagation is only possible for as far as the eye can see, the range of a space wave path is limited to at most some 45 to 60 km. If, as is usually the case, a longer route than this is required, then a number of 'hops' must be employed, as shown in Fig. 12.9. Microwave systems carrying large numbers of telephony channels use the space wave and are often known as **radio relay** systems.

Communication satellite

The principle of a communication satellite system is shown in Fig. 12.10. Carrier frequencies in the SHF band are used, one frequency for the up-path and another for the down-path in both directions of transmission, i.e. a total of four different frequencies. The ionosphere has negligible effect upon the radio waves and the waves travel in straight lines between each ground aerial and the satellite.

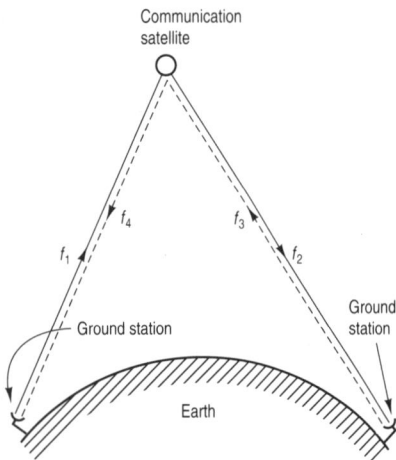

Fig. 12.10 Communication satellite system.

Radio receivers

The functions of a radio receiver are: (a) to select the wanted signal from all those signals picked up by the receiving aerial; (b) to demodulate, or detect, the wanted signal to extract the intelligence that has been modulated onto the carrier; and (c) to amplify the signal so that an output of sufficient power to operate the loudspeaker, or to apply to a telephone line, or whatever, is obtained.

Radio receivers all employ the 'superheterodyne' principle. The basic block diagram of a superheterodyne radio receiver is shown in Fig. 12.11. The wanted signal is selected by the radio-frequency circuit (which may include amplification) and applied to a circuit known as either a frequency-changer or a mixer. In the mixer circuit the wanted signal is mixed with the voltage produced by a local oscillator and the mixing process produces signals at frequencies equal to both the sum and the difference of the frequencies of the two signals. The output of the mixer is applied to an **intermediate-frequency amplifier** and this circuit selects and amplifies only the difference-frequency component. The output of the intermediate-frequency amplifier is then applied to the detector and here the baseband signal is recovered from the modulated wave. Lastly, the baseband signal is applied to an audio-frequency power amplifier to amplify the signal to the required power level.

Radio systems

The VLF and LF bands are used for very long distance narrow-band radio-telegraphy and radio-telephony systems, some navigation systems, and a few LF sound broadcasts. The medium waveband is mainly used for sound broadcasting but there are also some maritime services. The HF band is used for international radio-telephony systems and for some marine and aero systems. The other frequency bands are used for such services as sound and television broadcasting, various mobile systems, and to carry multi-channel telephony systems.

Figure 12.12 shows the essentials of a long-wave or medium-wave sound-broadcasting system. The information, speech, or music to be transmitted is produced live in a studio, or is obtained from an outside location, or has been previously recorded and is played back from a tape deck. The information signal is amplified before it is passed over a circuit, which may be either a land line or a radio link, to the

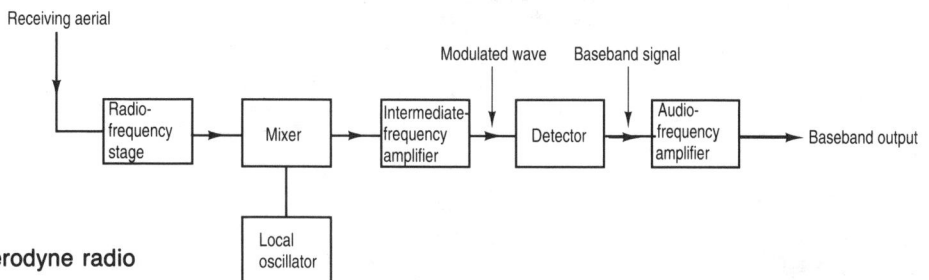

Fig. 12.11 Superheterodyne radio receiver.

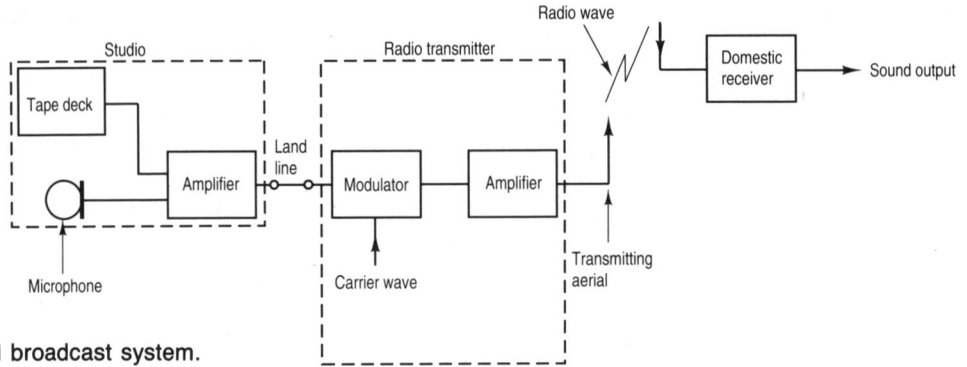

Fig. 12.12 Sound broadcast system.

radio transmitter. Here the baseband signal amplitude modulates a carrier whose frequency is the one allocated to that particular radio station. The amplitude-modulated wave is then amplified before it is applied to a transmitting aerial and radiated into the atmosphere. The frequencies used for sound broadcasting are such that the mode of propagation used is the ground wave. Programmes radiated at the lower frequencies in the medium waveband need only a single transmitter to provide coverage of the UK, but at the higher end of the band, where the range of the ground wave is smaller, it becomes necessary to have a number of transmitters situated in various parts of the country. The transmitted radio signal is picked up by all the receiving aerials in the service area of a transmitter and is selected by all the radio receivers tuned to that station.

The sky wave is used for some long-distance sound-broadcasting services, such as the BBC World Service, but its main use is for point-to-point radio telephony. The block diagram of an international radio-telephony system is given in Fig. 12.13. Different frequencies are used for each direction of transmission, otherwise it would be difficult to prevent a signal transmitted by a transmitter being received by its

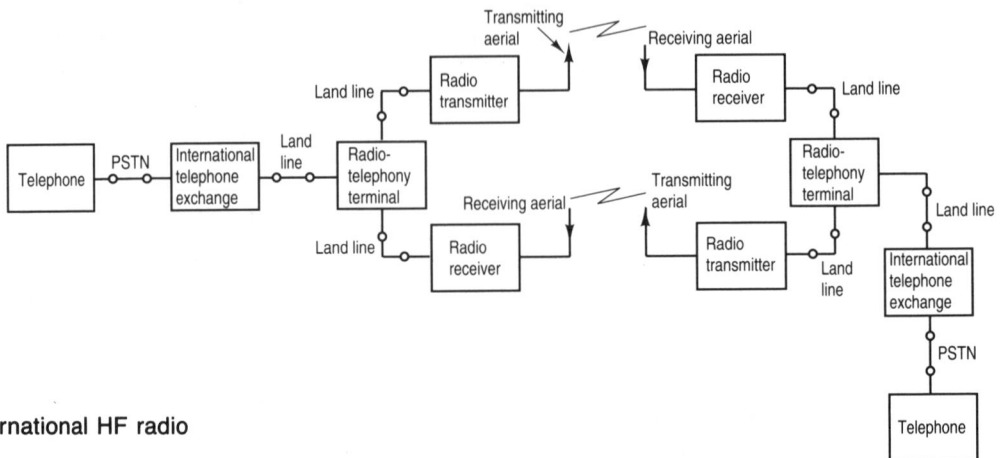

Fig. 12.13 International HF radio system.

associated receiver. The vast majority of international telephone calls are routed over multi-channel telephony systems that, in turn, are routed over either coaxial or fibre optic cable, or a terrestrial radio-relay system, or a communication satellite system, or some combination thereof. However, HF links are still used as important stand-by circuits and to provide communication over infrequently used routes. A call is routed over the PSTN to an international telephone exchange and thence to a radio-telephony terminal. At this terminal the call is connected to a radio transmitter and is then transmitted to the destination country.

Land-mobile radio systems are widely employed today by many different kinds of services, such as (a) **emergency** (ambulance, fire, and police), (b) **public utilities** (gas, electricity, and water), and (c) **private** (delivery vans, taxis, etc.). These systems are usually said to be **private land-mobile radio** (PMR) systems since they do not have access to the PSTN. Frequencies in the VHF and UHF bands are used, with different frequencies being allocated to different services. Other land-mobile systems exist that do have access to the PSTN, and these use a technique known as **cellular radio**. In all land-mobile systems at least one base station is used to communicate with the mobile stations.

PMR systems

The basic principle of any two-way land-mobile radio system is shown in Fig. 12.14. Either amplitude modulation or frequency modulation is employed and different carrier frequencies are used for each direction of transmission. An individual mobile transceiver is contacted by the base station by transmitting the unique code that identifies that mobile. When a transceiver recognizes that its code is being

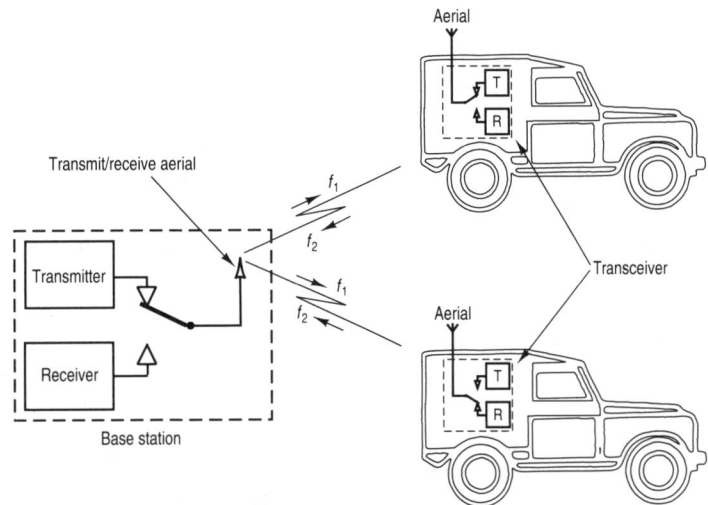

Fig. 12.14 Land-mobile radio system.

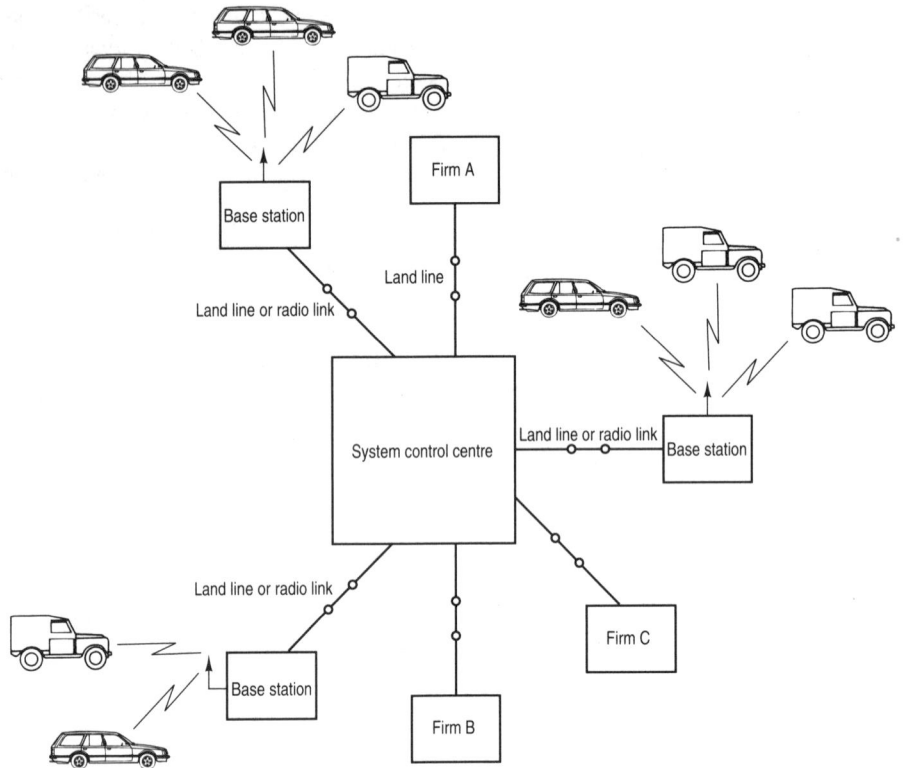

Fig. 12.15 Multi-user PMR system.

transmitted it will automatically answer the call. The base station then allocates a free channel to the call and conversation can begin. When a mobile wishes to make a call the transceiver searches the allocated frequency band for a free channel and seizes the first one it finds. The mobile then uses this channel to contact the base station.

A land-mobile radio system can be used by just one organization or it may be shared by several different companies. The basic arrangement of a multi-user system is shown in Fig. 12.15. The controlling station of each organization is connected, by land line or by point-to-point radio link, to the system control centre, and this centre, in turn, has land or radio links to several base stations scattered around the service area. The mobiles of each organization are able to communicate with the nearest base station and thence, via the system control centre, with their own controlling station.

Cellular radio

A cellular radio system provides access to the PSTN for mobile users over the whole of the service area. The service area is divided up into a large number of **cells** each of which has its own base station. The cells are grouped together in more or less hexagonal clusters and

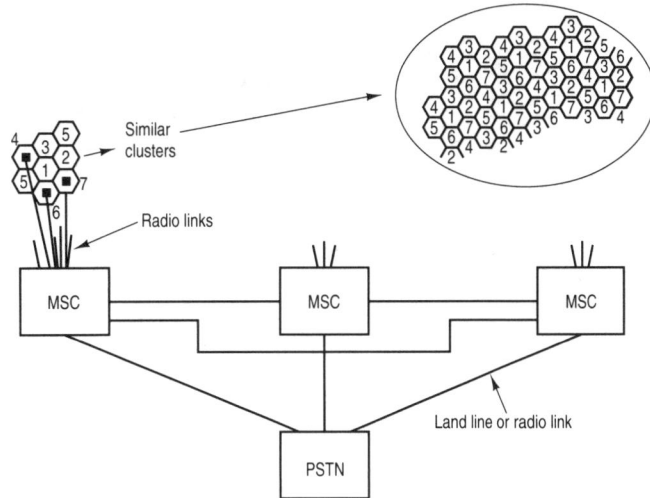

Fig. 12.16 Cellular radio.

each cluster is allocated a number of channels on different carrier frequencies. Each base station is connected by a land line to the nearest **mobile switching centre** (MSC) and all the MSCs are fully interconnected with each other *and* with the PSTN. The basic arrangement of the system is shown in Fig. 12.16. As each mobile transceiver moves around within the service area its location is continually tracked by the nearest MSC, and if it should move from one area to another, it is 'handed over' to the MSC in whose area it is now located.

When a mobile wishes to make a call it contacts the nearest MSC to its current location and is allocated a channel; the required telephone number is then 'dialled' and the call is connected to another mobile or to a number on the PSTN. If, while a call is in progress, the mobile moves into another cell the call is automatically taken over by the new base station. The handover takes place so rapidly that the user is unaware of its occurrence.

The earlier cellular radio system were analogue systems, but digital systems are increasingly being used.

Radio-paging

A radio-paging system is used to alert users that they are wanted by their office. Every pager is allocated a unique code and when this code is transmitted by the control station the pager recognizes its code and emits a 'bleep'. The call is also received by all other pagers in the system but since their code is not being received they do not respond to the call.

In the UK paging system there are 400 transmitters to cover 40 paging zones, each of which is polled by a number of computers.

The paging service uses two VHF channels, one of which operates at 153.125 MHz while the other operates at 153.175 MHz. Each pager works on one of the channels only, but all the transmitters can use both channels. The capacity of each channel is sufficient to serve one million pagers.

13 Television

A television system is concerned with the transmission of moving pictures and may be either **monochrome** (black and white) or **colour**. A colour television system is designed to be compatible with the earlier monochrome system, which means that the colour system is able to produce a monochrome picture on the screen of a black and white television receiver when that receiver is only receiving a colour signal. Reverse compatibility is also provided: a colour television receiver is able to reproduce a black and white picture when it is receiving a monochrome picture signal. To make compatibility possible the colour television system transmits monochrome information, known as the **luminance signal**, to which colour information, known as the **chrominance signal**, is added. A monochrome television receiver does not respond to the chrominance signal.

A television system uses a television camera to convert the light energy produced by an image into the corresponding electrical signal. The electrical signal is then transmitted, usually over a land line, to a television-transmitting station. Here the signals amplitude modulate a carrier frequency to position the picture signal in the part of the UHF band that has been allocated to that station. The associated sound signal frequency modulates another UHF carrier frequency to position the sound signal alongside the picture signal. The picture and sound signals are combined together and the composite signal is radiated into the atmosphere. The radiated signal is picked up by all the television receiving aerials located within the service area of the transmitter and is passed on to the associated television receivers. Each television-receiver selects the wanted programme signal from all those signals present at its aerial input terminal and amplifies it. The sound and picture signals are then separated from one another and separately processed. The processed picture signal is applied to a cathode ray tube to produce the wanted picture, and the sound signal is applied to a loudspeaker to produce the accompanying sound.

The basic block diagram of a television system is shown in Fig. 13.1.

Scanning

Television is concerned with the transmission of moving pictures. To give the viewer an impression of movement a succession of still pictures are shown, with each picture being slightly changed from

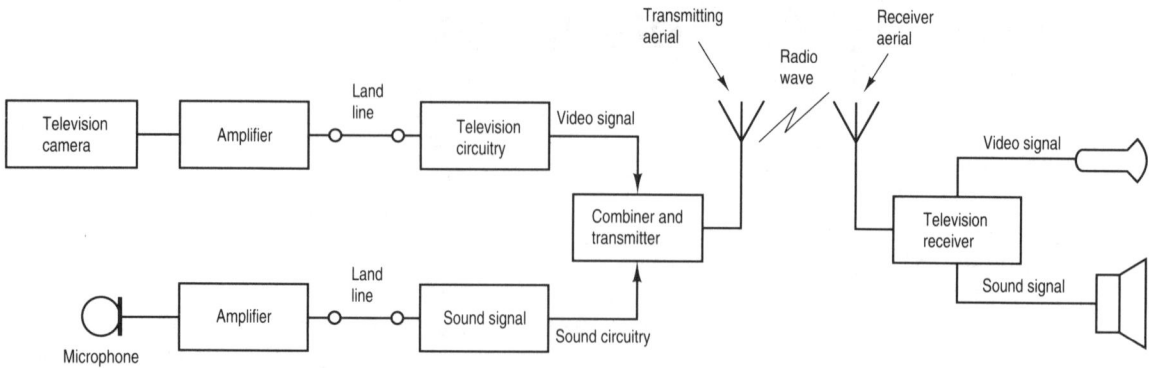

Fig. 13.1 Television system.

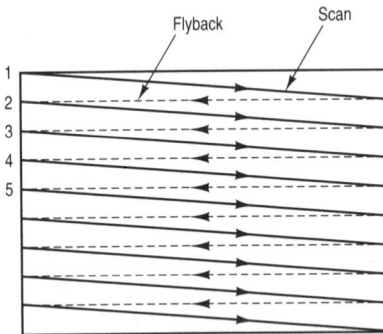

Fig. 13.2 Scanning.

the still picture that immediately preceded it. Provided the still pictures are shown rapidly enough the 'persistence of vision' of the human eye makes the viewer believe that he is seeing a moving picture.

The television camera uses a lens system to focus the light image to be televised onto a light-sensitive device known as an **area charge-coupled device** (CCD). The light-sensitive area is divided up into a large number of very small areas known as **pixels**, which are arranged in 625 horizontal lines. According to the focused light image each pixel will have a particular brightness which ranges, for a monochrome camera, from pure white, through all the possible shades of grey, to black. To transmit the brightness information which represents the light image the horizontal lines are scanned from left to right, as shown in Fig. 13.2. Once a line has been scanned the scan is rapidly moved back to the left-hand side of the picture — this is known as the **flyback** — and down slightly to the beginning of the next line, which is then scanned, and so on. In this way all the horizontal lines of pixels are scanned one after another. When the scan has reached the bottom of the picture it is rapidly returned to the top of the picture to repeat the scanning process with line 1.

As each pixel is scanned an electrical signal whose magnitude is directly proportional to the brightness of the pixel is outputted. The total electrical signal generated represents the detail of the light image to be televised, and is called the **picture signal**. The picture signal is an analogue waveform that has a maximum value of $+0.7$ V (peak white) and a minimum value of 0 V (black).

To avoid an annoying flicker being visible on the viewed picture it is necessary to show the still pictures at a rate of at least about 50 pictures per second. If, however, 50 pictures per second were to be transmitted, a very wide bandwidth would be required by the system and this would be expensive to provide. To reduce the necessary bandwidth a technique known as **interlaced scanning** is always employed in television broadcast systems. [Note, however, that the more expensive monitors used with computers do not employ interlaced scanning.] Interlaced scanning means that the odd-numbered

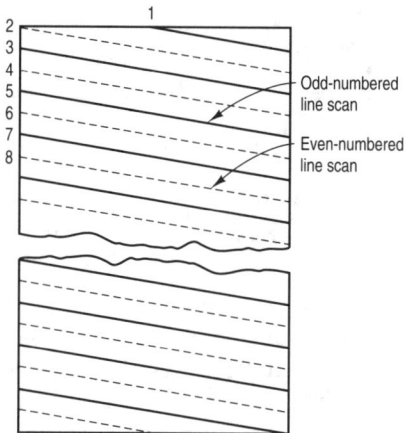

Fig. 13.3 Interlaced scanning.

lines (1, 3, 5, 7, etc.) are scanned first, starting half-way through line 1 as shown in Fig. 13.3. When the bottom of the picture is reached the scan is rapidly moved back to the top of the picture to start scanning the even-numbered lines (2, 4, 6, etc.). The last even-numbered line is scanned to half-way across the line before the scan moves back to start again mid-way along line 1. The total number of lines used in the UK system is 625.

Each complete scan of odd lines or even lines is known as a **field**, and hence each complete picture is made up of two fields. Each field is scanned 25 times per second and the effect is as though the picture has been scanned 50 times per second.

Synchronization pulses

At the television receiver the screen of the cathode-ray tube must be scanned in exactly the same way and with exactly the same speed as the television camera has scanned the light image. To ensure that the two scans are exactly in step it is necessary to add synchronizing (synch) pulses to the picture signal. The combination of the picture signal and the added synch pulses is known as the **video signal**. Both line and field synch pulses are used. Although there are 625 lines altogether, 25 lines per field are suppressed, i.e. 50 lines per picture. This means that each visible picture is made up from 574 complete lines plus two half-lines. Some of the suppressed lines are employed to carry the field synch pulses while others carry Teletext services. The synch pulses are at a voltage of -0.3 V and are therefore 'blacker than black', and so these pulses are not visible on the screen of the television receiver. The bandwidth occupied by the video signal is 0 to 5.5 MHz. A typical video signal is shown in Fig. 13.4.

The video waveform is used to amplitude modulate a UHF carrier but to reduce the bandwidth required most of the lower sideband is not transmitted. The suppression of most of one sideband is a system

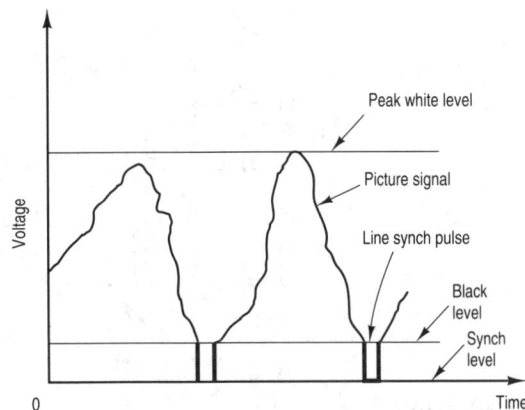

Fig. 13.4 Video signal.

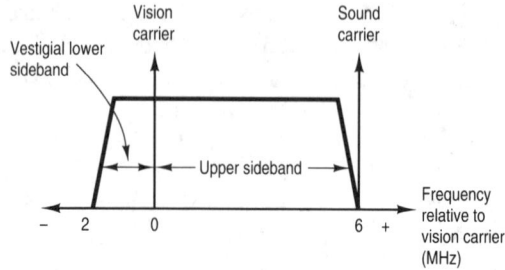

Fig. 13.5 VSBAM television signal.

known as **vestigial sideband amplitude modulation** (VSBAM), and is shown in Fig. 13.5. In the UK system negative modulation is employed; this means that the peak white signal is represented by the minimum voltage and a black signal by the maximum picture voltage, with the synch pulses at an even higher positive voltage.

Monochrome television

A monochrome television system employs a black-and-white television camera and a black-and-white television receiver. A monochrome camera uses an area CCD to convert incident light into an electrical signal. The CCD pick-up or imager is a LSI device that contains a very large number of photo-diodes ($454 \times 582 = 264\,228$, but only 418×575 are used for the picture area, and the remainder are kept at black level), plus 455 shift registers. [A shift register is a short-term storage device in which its stored data is moved one place to the right, or left, for each clock pulse.] The basic arrangement of the CCD pick-up device is shown in Fig. 13.6. The horizontal rows of photo-diodes correspond with the television scanning lines. The light image to be televised is focused onto the surface of the CCD by a glass lens system. A charge is developed in each photo-diode

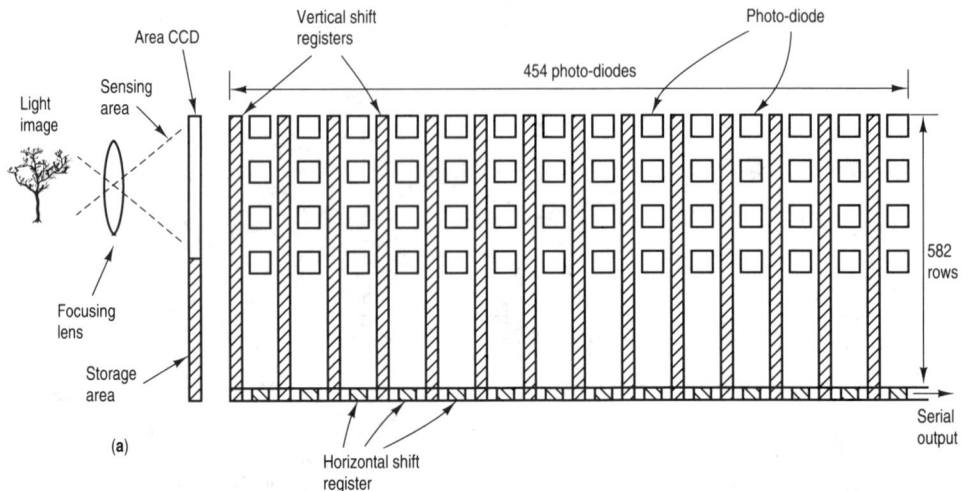

Fig. 13.6 CCD pick-up device.

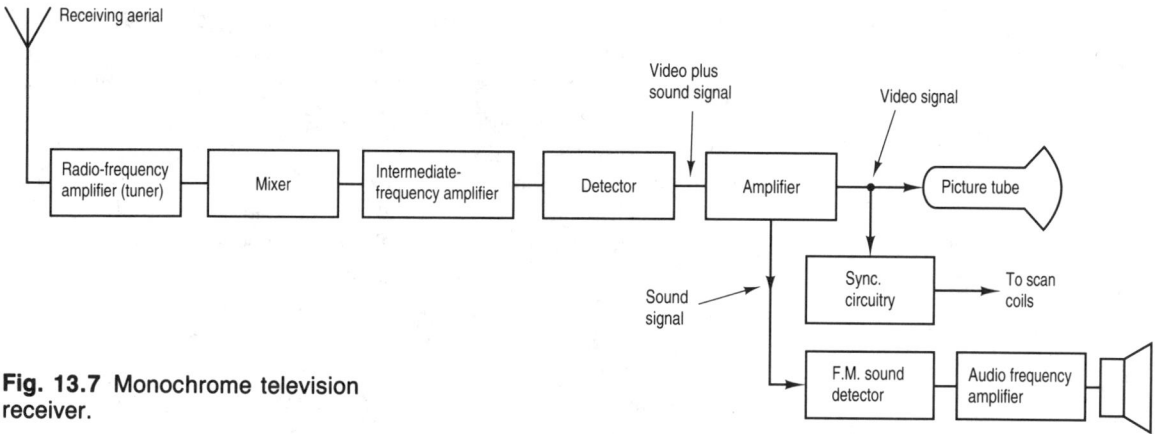

Fig. 13.7 Monochrome television receiver.

which is directly proportional to the intensity of the light incident upon it. During the vertical blanking period the entire charge representing the image is transferred from the sensing area to the storage area by stepping the charges downwards using the vertical shift registers. The picture signal is then read out in the way described on p. 000 for a colour camera.

The basic arrangement of a monochrome television receiver is shown in Fig. 13.7. The television signal picked up by the receiving aerial is selected from all those other signals that are also picked up, and the signal is amplified and demodulated to recover the baseband sound and video signals. The synch pulses are removed from the video signal and are used to synchronize the **timebase** in the receiver that controls the scanning of the cathode ray tube. The picture signal is applied to the tube to produce the wanted visible picture and the sound signal is applied to the loudspeaker.

Figure 13.8 gives a simple cross-section of a monochrome television

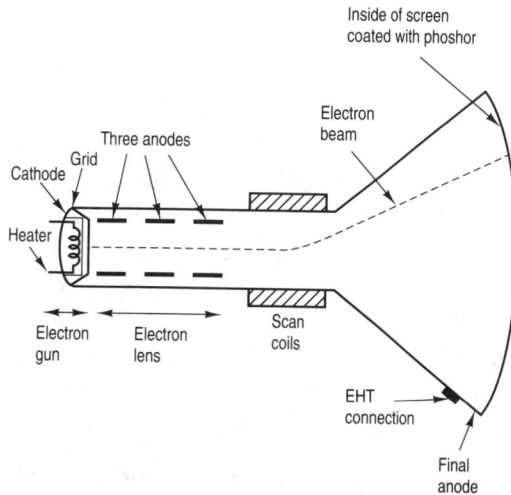

Fig. 13.8 Monochrome television receiver tube.

cathode ray tube. The cathode is surrounded by a cylindrical 'grid' which is held at a negative voltage. The cathode is heated to a high temperature by passing a current through its heater and this causes the cathode to emit large numbers of electrons. The grid controls the number of the emitted electrons that are able to leave the **electronic gun** and hence controls the brightness of the displayed picture. The emitted electrons are attracted towards the anode assembly which is held at a high positive voltage. The three anodes act as an electronic lens that both accelerates the emitted electrons and focuses them on to the screen. The inside and outside of the flared neck of the tube is coated with a substance known as aquadag; the inside coating is connected to a very high positive voltage, known as an **extremely high voltage** (EHT), and the outside coating is connected to the chassis of the receiver. The inside coating acts as a final anode that attracts and accelerates the emitted electrons to the screen. The overall effect is that a beam of electrons passes from the electron gun to the screen and here it is brought to a focus. The inside of the glass screen is coated with a fluorescent material which glows visibly when it is struck by the high-velocity electrons. The colour of the visible light is determined by the actual fluorescent material which is employed; for a monochrome receiver it is, of course, white.

When the electron beam is stationary a spot of white light is visible on the face of the screen and to obtain a display of some kind it is necessary to move this visible spot around the screen. Two sets of scan coils are positioned along the neck of the tube and these coils are used to deflect the electron beam. Line scan coils give deflection in the horizontal direction, and field scan coils give vertical deflection. By passing currents simultaneously through both sets of scan coils the electron beam, and hence the visible spot of light, can be deflected to any part of the screen.

Television receivers are commonly referred to in terms of their physical dimensions and this is usually quoted in terms of the diagonal measurement in centimetres of the visible picture (see Fig. 13.9). The **aspect ratio** of a tube is the ratio of its width to its height and this ratio is normally equal to 4/3.

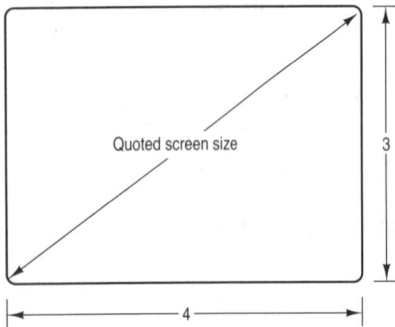

Fig. 13.9 Dimensions of a television receiver.

Example 13.1

A television tube gives a picture that has a diagonal dimension of 51 cm and an aspect ratio of 4/3. Calculate its height and its width.

Solution
$51^2 = h^2 + w^2$; $w = 4h/3$
$51^2 = 16h^2/9 + h^2 = 25h^2/9$
$h = 153/5 = 30.6$ cm. (*Ans.*)
$w = 4 \times 30.6/3 = 40.8$ cm. (*Ans.*)

Colour information

The three **primary colours** in radiated light are blue, green, and red. All other colours are obtained by adding some particular proportions of two of the primary colours together. The most important examples of additive mixing are:

blue + red = magenta (a bluish red)
green + red = yellow
blue + green = cyan (a bluish green).

If the three primary colours are added together in the proportions of 11% blue, 59% green, and 30% red, white light is obtained. The different proportions are necessary because the human eye is not equally sensitive over the whole of the colour spectrum. Cyan, magenta, and yellow are known as the three **complementary colours**. If any two of the complementary colours are added together in the correct proportions white light is obtained.

A colour signal is described by its **luminance**, its **hue**, and its **saturation**. Luminance is the brightness of a colour. To obtain white light the luminance signal, usually denoted by Y, must be equal to the sum of the primary colours in the quoted proportions. Thus

$$Y = 0.3R + 0.59G + 0.11B \qquad (13.1)$$

The term 'hue' describes the colour which corresponds to the dominant wavelength of a band of colours in the same group. The red hue, for example, includes all shades of red and pink. Lastly, the term 'saturation' indicates the degree to which a colour has been diluted with white light. If a colour has zero white light content it is said to be 100% saturated, whereas zero saturation indicates white light.

Example 13.2

A colour television camera has a maximum output voltage of 1 V.
1. Calculate the voltage of the luminance signal when the televised scene is (a) all white, (b) all black, (c) all red.
2. What will be the output voltage when the brightness of the white light is reduced by 50%?

Solution
1. (a) $Y = 0.3 \times 1 + 0.11 \times 1 + 0.59 \times 1 = 1$ V. (*Ans.*)
    (b) $Y = 0.3 \times 0 + 0.11 \times 0 + 0.59 \times 0 = 0$ V. (*Ans.*)
    (c) $Y = 0.3 \times 1 + 0.11 \times 0 + 0.59 \times 0 = 0.3$ V. (*Ans.*)
2. (d) $Y = 0.3 \times 0.5 + 0.11 \times 0.5 + 0.59 \times 0.5 = 0.5$ V. (*Ans.*)

Colour difference signals

The problem of achieving compatibility between colour and monochrome television systems is solved by transmitting the picture signal in two parts known as the **luminance signal** and the **chrominance signal**. The basic circuitry of a colour television receiver is shown in Fig. 13.10. The luminance signal is used to control the

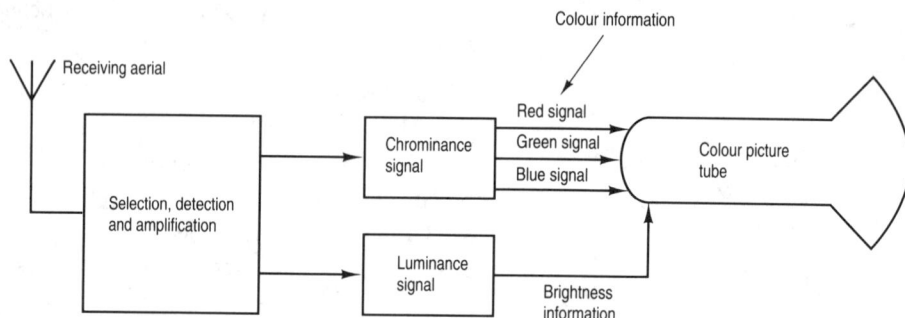

Fig. 13.10 Basic block diagram of a colour television receiver.

brightness of the displayed picture and is the only signal used by a monochrome receiver. The chrominance signal controls both the hue and the saturation of the colours in the displayed picture.

The chrominance signal consists of two separate components, known respectively as the red colour-difference signal and the blue colour-difference signal. A green colour-difference signal is also needed at the receiver but since it can be generated from the other two colour-difference signals it need not be transmitted. Not transmitting the green colour-difference signal reduces the bandwidth occupied by the chrominance signal.

The red colour-difference signal is $(R - Y)$ and the blue colour-difference signal is $(B - Y)$. The R and B signals can be recovered at the receiver from the colour-difference signals by merely adding the Y signal. Thus:

$$(R - Y) + Y = R \quad \text{and} \quad (B - Y) + Y = B.$$

Recovery of the green colour-difference signal is somewhat more complex. The luminance signal Y is

$$Y = 0.3R + 0.11B + 0.59G$$

and hence

$$\begin{aligned} 0.59G &= Y - 0.3R - 0.11B \\ &= Y - 0.3[(R - Y) + Y] - 0.11[(B - Y) \\ &\quad + Y] \\ &= 0.59Y - 0.3(R - Y) - 0.11(B - Y) \\ 0.59(G - Y) &= -0.3(R - Y) - 0.11(B - Y) \end{aligned}$$

or

$$G - Y = -0.51(R - Y) - 0.19(B - Y) \qquad (13.2)$$

The minus signs merely denote a reversal in phase.

This means that the green colour-difference signal can be obtained by adding 0.51 times the red colour-difference signal to 0.19 times the blue colour-difference signal and then inverting the sign of the sum. For either white, or for black, each of the three colour-difference signals is equal to zero.

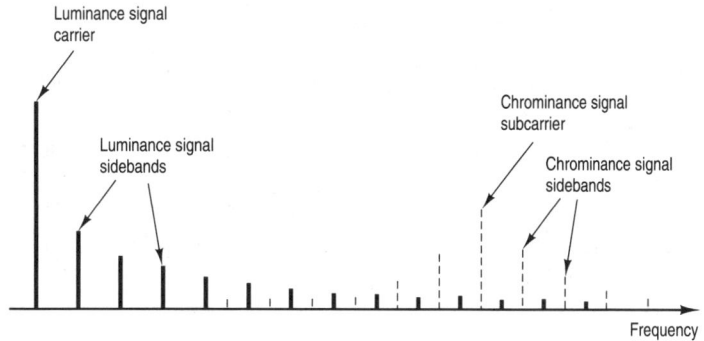

Fig. 13.11 Spectrum diagram of a colour television signal.

The two colour-difference signals, $(R - Y)$ and $(B - Y)$, are transmitted using quadrature amplitude modulation (QAM) of a subcarrier. The spectrum diagram is shown in Fig. 13.11 from which it can be seen that the sidebands of the chrominance signal are interleaved in the gaps which exist between the sidebands of the luminance signal. This method of transmitting the colour-difference signals ensures that colour information can be added to the luminance signal without any increase in the occupied bandwidth.

Colour television

A colour television system uses the same basic arrangement as a monochrome television system except that, of course, a colour camera and colour receiver are employed. A colour television camera uses an area CCD like a monochrome camera but the face of the image area is covered with a striped colour filter. Because the light transmission of red and blue light filters is low, the colour filter has cyan, green, magenta, and yellow elements that pass these colours only. Figure 13.12 shows the layout of the colour filter that is placed on top of the CCD surface. A cyan filter blocks red light but passes both blue light and green light; a magenta filter passes both red and blue light but not green light; and a yellow filter blocks blue light but passes both red and green light. In alternate lines the output signal is $(G + Cy)$, $(Mg + Ye)$, $(G + Cy)$, $(Mg + Ye)$ and so on, and $(Mg + Cy)$, $(G + Ye)$, $(Mg + Cy)$, $(G + Ye)$ and so on.

$$(G + Cy) = G + B + G = 2G + B$$

and

$$(Mg + Ye) = R + B + R + G = 2R + B + G$$

Hence in even-numbered lines,

$$Cy + G + Ye + Mg = 3G + 2R + 2B$$

which is a luminance component. Similarly in odd-numbered lines,

$$Mg + Cy + Ye + G = 2R + 2B + 3G$$

which is another luminance component. Also, in even-numbered lines

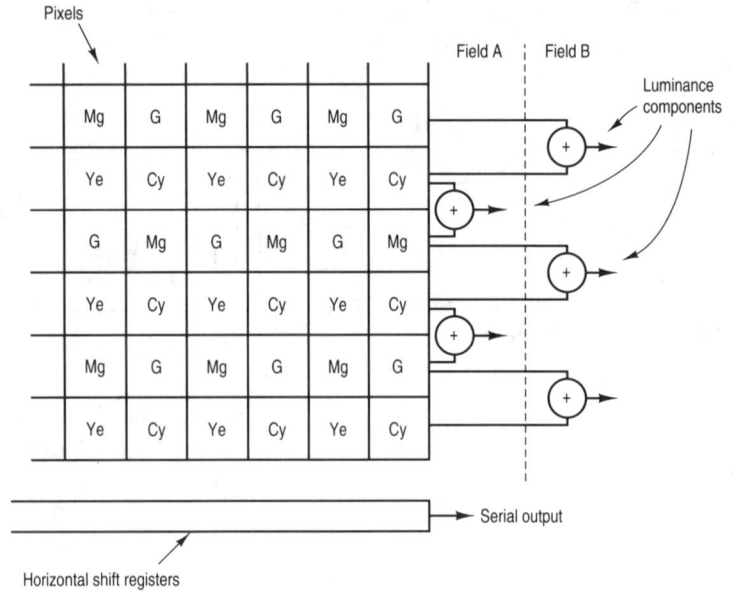

Fig. 13.12 CCD pick-up device colour filters.

$$(Mg + Ye) - (G + Cy) = 2R - G$$

and, in odd-numbered lines,

$$(Mg + Cy) - (G + Ye) = 2B - G.$$

The colour television camera usually employs line transfer of the photo-diode charges to the horizontal shift registers. During the vertical blanking period the charges on each of the photo-diodes is simultaneously shifted into a vertical shift register. These shift registers then work together during successive line-blanking periods to move these charges downwards, one row at a time, until a row reaches the bottom level. Then the charge is transferred into a horizontal shift register. The charges held in the horizontal shift registers are then moved out serially to give the picture signal.

Figure 13.13 shows, using a very simplified block diagram, how

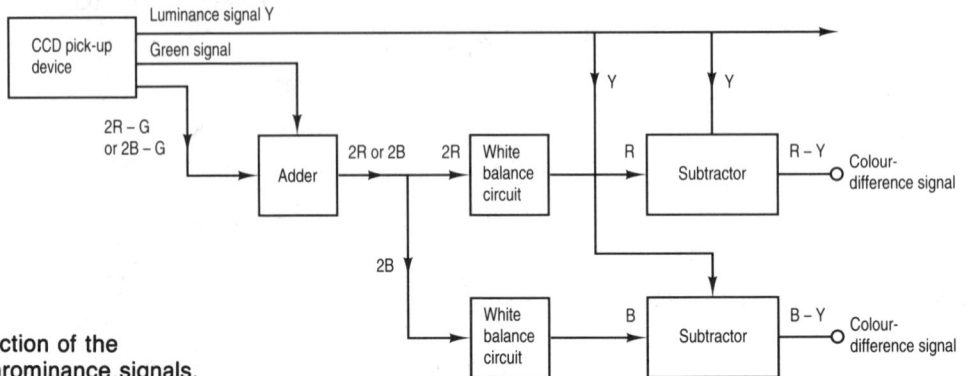

Fig. 13.13 Production of the luminance and chrominance signals.

the colour difference signals (R − Y) and (B − Y) are produced in a colour television camera. The G output of the CCD is added to the (R − Y) or (B − Y) component to remove the G component to give either 2R or 2B. The colour signals R and B are then applied to a subtracting circuit together with the luminance signal, and the outputs of the two subtractors are the (R − Y) and the (B − Y) colour-difference signals.

The two colour-difference signals modulate a sub-carrier using QAM and the resultant modulated signals are added together before being combined with the luminance signal. The composite video signal is then applied to a VSBAM modulator to position the signal in the allocated part of the UHF band. The basic arrangement used is shown in Fig. 13.14.

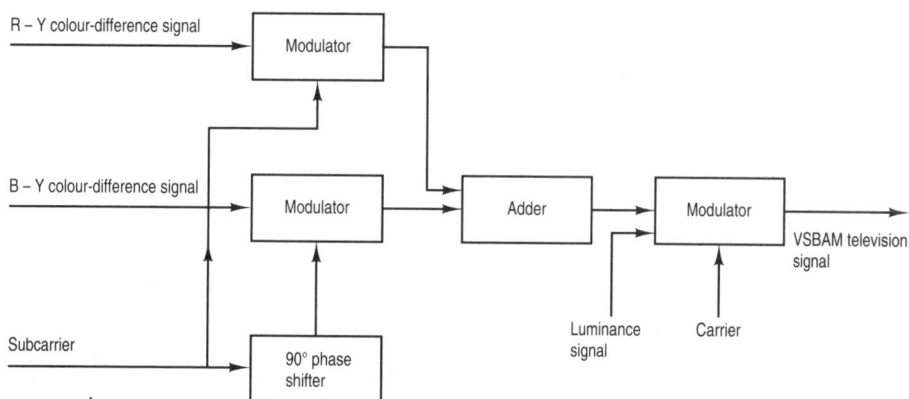

Fig. 13.14 How luminance and chrominance signals are combined in a colour television camera.

Colour tubes

A colour cathode-ray tube uses three separate electron guns, one for each primary colour, to direct three different electron beams onto the screen. The inside of the screen is coated with three different fluorescent materials, or phosphors, each of which glows with one of the primary colours when struck by high-velocity electrons. The three electron guns are mounted side by side to produce three separate electron beams (red, green, and blue). The phosphors are arranged in a large number of groups of three vertical stripes. The basic construction is shown in Fig. 13.15. It is necessary to have some kind of arrangement that will ensure that, as the screen is scanned, the red, blue, and green electron beams can only strike their respective phosphors. This requirement is satisfied by the insertion of a **shadow mask** immediately behind the screen; the shadow mask has a large number of vertical slots cut into it and is positioned so that it allows each electron beam to pass through a slot which allows it to strike only its own phosphor. When a phosphor is struck by a beam it glows visibly with the colour produced by that phosphor.

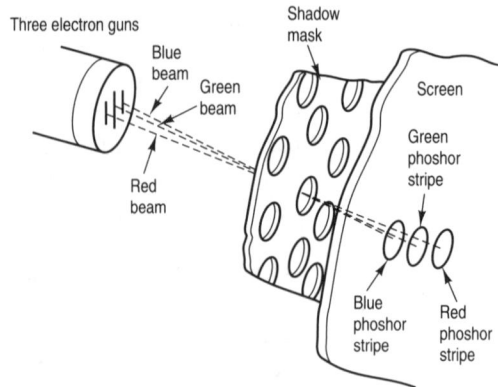

Fig. 13.15 Principle of a colour television tube.

NICAM

NICAM (near instantaneous companded audio multiplex) is a relatively new sound system that is used in modern television receivers. It is a digital system that provides high-quality stereo sound signals and is also able to carry data information. NICAM has only become possible since complex integrated circuits have become available.

Teletext

Teletext is a form of data transmission that is provided by the television broadcast companies in addition to the normal television programmes. In the UK there are two teletext services: one service, provided by the BBC, is known as Ceefax; the other service, provided by the commercial television companies, is known as Teletext.

Some of the horizontal lines that are not used for the transmission of picture information are used to transmit the data signals. The number of lines used for picture signals is 625 but only 575 of them are employed to produce a visible picture. The remaining lines are blanked out to carry field synch pulses and to give time for the field flyback to return to the top of the picture. Some of these blanked out lines are used for the transmission of the Teletext data signals. Because the data pulses are 'blacker than black' they are not visible on the television screen. An integrated circuit decoder is needed to store the incoming data signals and convert them into text that will be visible on-screen. In the UK television system 16 lines per field (7 to 22 on even-numbered fields and 320 to 335 on odd-numbered fields) are used for the Teletext service. The data rate which is employed is 6.9375 Mbits/s.

The displayed information is grouped into **pages** which the viewer is able to access by keying in the number of the required page using a remote control. Each page of data contains up to 24 pages of text and each row may contain up to 40 alphanumeric characters.

Viewdata

Viewdata is the name given to a service that provides computer-based information via a telephone line and uses either a television receiver or a VDU to display the data. The British Telecom version of viewdata is known as Prestel. The data called up by a user is displayed using a similar page format to that used with the teletext services. When the main index is called up a menu of one of many categories of information, or of services, is displayed and one of them may be selected. The subject index is then displayed and the user can then make a more specific choice. User selection continues until the wanted page of information has been found and is displayed on-screen. Unlike Teletext, two-way communication is possible and this allow queries to be made and to be answered. A user is able to communicate with such organizations as banks, insurance companies, large shops, and travel agents.

Exercises

Chapter 1: Information technology

1.1 Explain the meanings of the terms (a) information, (b) message transmission, (c) quality of information, and (d) information technology. For each give suitable examples.

1.2 Make a list of five practical examples of information technology that might be found in a modern office. For each item listed explain its function.

1.3 Explain the principles of operation of (a) Telex and (b) electronic mail. When and why might each be used? What is meant by a bulletin board?

1.4 A big problem in all offices is the efficient storage and retrieval of documents. List five different ways in which documents may be stored and discuss their relative merits.

1.5 Why are bar codes commonly used to identify products in shops today? Briefly explain how a bar coding works. What is meant by a PoS terminal and what advantages does its use bring to a large shop or department store? Credit card terminals are also commonly employed today. Explain briefly how the system works and state the advantages to the retailer. Draw the block diagram of a EFTPoS system and outline its operation.

1.6 Why are money transactions often carried out using electronic means in modern society? Explain the basic concepts of systems that (a) allow a person to obtain money from a cash dispenser, (b) allow a person to pay for goods using a credit card, and (c) allow a firm to pay salaries directly into an employee's bank account.

1.7 What is meant by an 'on-line' database? Give some examples.

1.8 Robots are able to perform many tasks perfectly adequately but there are a number of jobs for which robots will probably never be used. Give some examples of each. What are the differences between CAD and CAM?

1.9 An airline booking system includes a database of flight bookings. A travel agent can access the database to book seats for customers. Describe the information that the agent must input to the database to check if seats are available and the information that should be returned to the agent. What information will the airline want to be able to get from the database?

1.10 Discuss the various applications for digital computers in the fields of commerce and retailing.

Chapter 2: Digital computers

2.1 Draw the basic block diagram of a digital computer and describe the function of each of the blocks shown in the running of a program. Explain, with the aid of a flow chart, the meaning of the term 'fetch—execute cycle'.

2.2 Explain the functions of (a) the accumulator, (b) the program counter, and (c) the instruction decoder in a digital computer.

2.3 A digital computer may be either a user-programmable or a stored-program machine. (a) Explain the difference between the two kinds of machine. (b) Briefly discuss the advantages and disadvantages of stored-program machines. Explain the difference between real-time processing and batch processing of data.

2.4 Digital computers are generally categorized as being either mainframe, mini-, or micro-, types although the boundaries between them are becoming increasingly blurred. Discuss the differences between these kinds of computer. Why are many microcomputers popularly known as personal computers? For what purposes is each type normally employed?

2.5 Explain the events that occur when a computer is first switched on. How is the user informed when program/data entry may begin? What is meant by (a) the instruction cycle and (b) the execution cycle?

2.6 Explain, with the aid of suitable diagrams, how a computer reads data out of, and writes data into, a selected memory location.

2.7 Draw a block diagram of a digital computer and use it to explain the use of the address, control, and data buses. What is meant by the term 'bus width'? If an address bus has a width of 32, how many memory locations is it able to access?

2.8 What is **cache memory**? Explain why it may be used (a) as a part of the main memory and (b) between the computer and the disk drive. How is the CPU of (a) a mainframe computer and (b) a PC provided?

2.9 Why must a program that is being run be held in the internal memory of the computer? Explain the differences between the internal memory ICs RAM and ROM. For what purposes is each employed? What is meant by the terms (a) volatile and (b) non-volatile, and which is RAM or ROM?

Chapter 3: Input/output devices

3.1 What is meant by a key-to-disk system for the input of data to a digital computer? What are the advantages to be gained by the use of such a system and when might it be employed?

3.2 Most digital computers are provided with both internal memory and backing storage. Explain the purpose of each form of memory. List the various kinds of backing storage that are commonly employed and state (a) their relative merits and (b) when and where each might be used.

3.3 Explain the difference between a hard disk and a floppy disk and give a sketch of one of them. Show how a disk may be organized into a number of tracks and sectors. What is meant by the packing density of a disk?

3.4 Magnetic tape is sometimes employed for backing storage in a computer system. What forms of tape storage are used and for what purposes? Explain the basic principle of either a microfilm or a microfiche storage system and say when it might be used.

3.5 When data is accessed from the backing storage of a computer system the access time depends upon whether parallel or serial access is employed. What is meant by 'parallel' and 'serial access' and when and where is each used?

3.6 Explain the meanings of the following terms that are used in conjunction with digital computer memory systems.
(a)   Access time, latency time, and cycle time.
(b)   Immediate access memory.
(c)   Volatile and non-volatile.
(d)   Destructive and non-destructive read-out.
(e)   Scratchpad memory.

3.7 Describe the function and basic principle of operation of a computer monitor. What is the difference between a monitor and a visual display unit (VDU)? How is the resolution of a monitor's screen quoted? What is the function of a video adaptor?

3.8 Describe the operation and typical applications of (a) MICR and (b) OCR. List their advantages and disadvantages.

3.9 Magnetic tape remains an important form of backing storage for computers even though its access time is relatively slow. (a) Give some applications where this is true. (b) Explain why the access time is much longer.

Chapter 4: Software

4.1 What is meant by (a) system software, (b) support or utility software, and (c) application software? Give two examples of each.

4.2 Some programs are always held in ROM within a computer. Why is this necessary and which programs may be so stored? List the functions of an operating system and say why it is the usual practice to load this program from disk immediately the computer is switched on. Name some operating systems in common use and say whether each is used in a mainframe or a personal computer.

4.3 What are the main differences between a 'batch-processing system' and a 'real-time system'. Give an example of an application that is best suited to each. Explain the principle of operation of (a) a multi-user system and (b) a multi-tasking system.

4.4 List the differences between an interpreter and a compiler. Why is the former often better suited to use with small computers? Give two examples of high-level languages that are (a) compiled and (b) interpreted.

4.5 What is meant by an algorithm? Write down an algorithm, in flow chart form if preferred, to find the memory location of the highest of a number of numeric items held in 20 successive memory locations.

4.6 How can a loop in a program be used to (a) handle multiple sets of data, and (b) repeat until a specified condition is met? A loop is often associated with a JUMP instruction; explain the difference between a conditional jump and an unconditional jump in a program and give an example of each.

4.7 One hundred small integer numbers which are stored in successive memory locations are to be added together and their sum is to be placed into the memory location immediately above the highest holding an integer number. Write a low-level program to carry out this task.

4.8 Explain the differences between machine code, assembly language, and high-level language programs. When and why must a program to convert a low-level or a high-level language into machine code be used? Give a name to the program used in each case. Why are subroutines often employed in a low-level language program? Subroutines can also be used in high-level programs but some languages allow the use of procedures instead. What is the advantage to be gained from the use of procedures?

4.9 Explain the purpose of a subroutine. State the instructions used to initiate and to terminate a subroutine. Write a subroutine to provide a time delay.

4.10 With the aid of a flow chart explain each of the following program instructions: (a) WHILE DO, (b) IF THEN ELSE, and (c) REPEAT UNTIL.

4.11 Describe the use of a graphical user interface (GUI) to simplify the operation of a computer. What is a mouse and why is one commonly used in conjunction with a GUI? What advantages are claimed for GUIs?

4.12 Discuss the use of databases and spreadsheets in modern offices. There are two main kinds of database; name them and discuss their relative merits. What is an integrated software package? Many integrated packages contain four different types of program; what are these most likely to be?

4.13 Draw a flow chart for the problem of making a cup of tea.

4.14 (a) What are the three main categories of computer instructions? For each category give three examples. (b) What is meant by a directive and why are they required in low-level programming?

[For Exercises 4.15–4.17 use a simple instruction set containing the instructions LOAD, STORE, ADD, DECREMENT, HALT, and CONDITIONAL and UNCONDITIONAL JUMP.]

4.15 Using the simple instruction set, write a program to subtract the numbers held in memory address ONE from the number stored at address TWO and then store the result at the address held in register X.

4.16 Write a program to find the sum of the numbers 1 to 150.

4.17 Write a program to find how many integer numbers, starting from 1, must be added together before their sum becomes larger than 255.

Chapter 5: IT in the office

5.1 Outline the basic characteristics of document preparation using (a) typewriters with no memory, (b) typewriters with limited memory, and (c) word processors. What are the component parts of a stand-alone word processor?

5.2 List the advantages of a word processor over a typewriter. List ten features that might be found in a word processor. Why may it be an advantage to use an integrated software package?

5.3 Explain what is meant by (a) a database and (b) a spreadsheet making clear how they differ from one another. Demonstrate the setting up of a spreadsheet by considering the spending of a week's housekeeping money in a typical household. Explain how a file in a database might be set up.

5.4 Describe briefly the way in which each of the following printers work: (a) daisy-wheel, (b) dot-matrix, and (c) ink-jet. Which of these printers is able to give carbon copies? List the relative advantages and disadvantages of these three printers.

5.5 Describe the operation of (a) a photo-copier and (b) a

laser printer. What are the main advantages and disadvantages of the laser printer compared to the other types that are available and why are they increasingly employed?

5.6 Discuss the main features of any word-processing package with which you are familiar. Give the steps that are taken in the preparation of a document and explain how the document is edited. Explain the meanings of the word processing terms 'justification', 'word wrap', 'tabs', and 'line pitch'. What is meant by mail merge?

5.7 The following terms are used in conjunction with databases. Explain the meaning of each term: (a) flat-file, (b) relational, (c) file, (d) record, and (e) field.

5.8 Derive a spreadsheet to determine the probable cost of a house that you hope to buy. The house will cost £x, and the total legal, surveying, and removal costs are £y. The spreadsheet should cater for different percentage deposits and for a mortgage rate that varies (hopefully!) only once per year. Make clear the formulae that are to be entered into the spreadsheet. Enter typical figures for all the variables and try to discover what would be the best financial deal.

5.9 Explain the need for data protection and distinguish between data privacy and data security. Give some typical examples to show how unauthorized access to data can be prevented.

Chapter 6: Analogue and digital signals

6.1 Give one example of an analogue device and one example of a digital device and say what each would be used for. What is meant by (a) bit, (b) bit rate, (c) bipolar digital signal, and (d) unipolar digital signal? When a system is controlled by a microprocessor the transducers used are most often analogue devices. A microprocessor can only deal with digital signals. Show how this difficulty is overcome in a practical system.

6.2 A computer is used to analyse the signals obtained from two analogue transducers and to output a numerical result to a digital display. Draw a block diagram of the system showing whether the signal at the input and output of each block is analogue or digital.

6.3 Explain what is meant by an integrated circuit (IC). What are the two main classifications of ICs and give one example of each. Digital ICs are further classified in terms of their complexity; state the four classes used and give one example from each.

6.4 What is meant by (a) an analogue and (b) a digital signal? Give two examples of each. Which kind of signal is produced by each of the following transducers: (a) telephone transmitter, (b) shaft encoder, (c) thermostat, (d) thermistor, (e) television camera?

6.5 A bipolar rectangular waveform has 3 V positive pulses of 1 ms duration and 2 V negative pulses of 2 ms duration. Draw the waveform and determine (a) its mark—space ratio, (b) its duty factor, and (c) its mean value.

6.6 Explain the difference between an analogue-to-digital converter and a digital-to-analogue converter. Draw a block diagram to show how each would be used in a computer-controlled process. A 12-bit DAC has a maximum analogue output voltage of 5 V. Show that its resolution is 1.22 mV.

Chapter 7: Microelectronic systems

7.1 Show, with the aid of a block diagram and a flow chart, how an automatic washing machine can be controlled by a microprocessor.

7.2 A hot drinks machine provides drinks when the correct money has been inserted and a control has been pressed. The machine uses five ingredients: coffee (C), milk (M), sugar (S), tea (T), and water (W). The drinks that are available are as follows:

(1) T + W; (2) T + W + M; (3) T + W + M + S; (4) C + W; (5) C + W + M; (6) C + W + M + S.

Fixed quantities of the ingredients are placed into a bowl via valves which open and shut under microprocessor control. Water must only be added after the other ingredients and after a plastic cup has been placed beneath the bowl ready to catch the made drink. Draw the flow chart for the system.

7.3 Draw a flow chart for a burglar alarm system in which four switches are used. Switch W is used to detect if a window is opened, switch D to detect if a door is opened, switch S is a sound detector, and switch F detects pressure on a floor.

7.4 There are two different methods used by a microprocessor to input and output data to/from a peripheral. Describe briefly these methods and state their relative merits. Name one microprocessor that uses each of the methods. Why is an interface IC required to connect a peripheral to a microprocessor?

7.5 Explain why a microcomputer may be regarded as a system. What is a transducer and what is its function? With the aid of a resistance-temperature characteristic explain how a thermistor may be used to measure temperature. How could the transducer be interfaced to a microprocessor?

7.6 A transducer produces an output voltage that is directly proportional to temperature. The transducer is to be interfaced to a microprocessor system that will be used to switch a heater on whenever the temperature falls below 15°C, and to switch the heater off when the temperature is above 20°C. Draw the block diagram of the system including all buses. Suggest a suitable transducer.

7.7 Each of the following items might be used in a microprocessor-controlled system. State the function of each:

(a) transducer, (b) analogue-to-digital converter, (c) digital-to-analogue converter, (c) visual display unit, (d) ROM, (e) backing store, and (f) assembler.

7.8 Draw a simplified block diagram of a typical microprocessor. Explain how the device operates to perform the instruction 'add the decimal number 12 to the number held in memory location $0350 and store the sum in location $0352'.

Chapter 8: Assembly language programming

8.1 A segment of a program is:

```
CLC
LDA #06
STA $0251
LDA #03
ADD $0251
END
```

When the program has been executed what does the accumulator hold? If the second and the fifth instructions are transposed and the program re-run, what will the accumulator then hold?

8.2 A microprocessor may be addressed in either of the three following modes: immediate, extended, and direct. Describe the characteristics of each mode and give two examples of the use of each.

8.3 A sequence of instructions that is stored in the memory locations 0310 to 0330 is to be executed three times. Write a suitable assembly language program before and after this block that will satisfy the requirement.

8.4 Write short programs for the 6809, 80386, and 68030 microprocessors to

(a) add the contents of location $0360 to the contents of location $0363 and store the result in location $0360;

(b) load one register with 3 and another register with 5, then add the two numbers together and store the result in location $0360;

(c) change over the contents of the memory locations $0360 and $0380;

(d) calculate the sum and the difference of the two numbers stored at locations $0360 and $0370 and then store the sum at location $0362 and the difference at location $0372.

8.5 Write a program for the 6809, 80386, and 68030 microprocessors that will check the two numbers stored in locations $0360 and $0361, and if the larger number is in location 0360 will interchange the contents.

8.6 Write a program for the 6809, 80486, and 68030 microprocessors that will store the integer numbers 5 to 15 in successive memory locations starting from $0400.

8.7 What is a subroutine and why is it used? State some examples of tasks that might well be included in subroutines. Write a program that will execute a subroutine residing at location $A200 five times.

8.8 A loop is required that should be terminated when a process has been carried out ten times. Explain three different ways in which this loop may be formed.

8.9 Write a program to implement the flow chart of an OR gate.

Chapter 9: Information transmission

9.1 A sinusoidal voltage has a peak voltage of 10 V and a frequency of 1 kHz. Draw two cycles of the waveform (a) to a time axis and (b) to a distance axis. Mark on the first figure the periodic time of the waveform and label it in milliseconds, and mark the second figure with the wavelength of the signal. Why can the wavelength not be calculated?

9.2 A rectangular bipolar wave has a bit rate of 500 bits/s. What is the duration of each bit? Why may the frequency of the wave vary even though the bit rate remains constant? If the average or mean value of the waveform is zero what will happen if the d.c. component is removed?

9.3 Draw a sinusoidal voltage waveform with a peak value of 12 V. Then draw, on the same axes, another sine wave at twice the frequency and of peak voltage 4 V. Thence plot the resultant waveform and comment on the result.

9.4 A unipolar rectangular voltage wave has pulses of 1 ms duration and a periodic time of 4 ms. Draw the spectrum diagram of the waveform.

9.5 List all of the frequency components, and their amplitudes, of (a) a square wave, (b) a triangular wave, if the peak voltage of each wave is 5 V. Why is it generally not possible to transmit all of these components in a practical system and what is the effect of not doing so?

9.6 What is meant by the term 'modulation' and why is it used in communication systems? Which of the parameters of a sinusoidal carrier wave are modulated for (a) amplitude modulation, (b) frequency modulation, and (c) phase modulation? Which modulation method is employed for sound broadcasting in (a) the medium wave and (b) the VHF band?

9.7 A 1 MHz carrier wave is amplitude modulated by a 0 to 4 kHz baseband signal. Explain why the bandwidth occupied by the modulated wave is twice as large as the baseband bandwidth. Draw the spectrum diagram of the modulated signal.

9.8 How can digital signals be represented by modulated voice-frequency signals? Why is such representation necessary? Quadrature amplitude modulation (QAM) is often used in modern data communication systems; what parameters of a sinusoidal carrier are then modulated?

Chapter 10: The public switched telephone network

10.1 Describe, using a simple block diagram, how a telephone customer on one exchange is connected to a customer on another exchange.

10.2 Explain, with the aid of diagrams, why it is impractical to consider directly connecting all the customers of a telephone network. Show how two customers may be connected together via one, or more, telephone exchanges.

10.3 Explain the need for supervisory signals, such as dial tone, ringing tone, and busy tone, in a switched telephone system.

10.4 Draw a 6-inlet, 4-outlet matrix switch and explain how it may be used to connect any inlet to any outlet. For this switch what is (a) the number of crosspoints, and (b) the maximum number of simultaneous calls that can be carried?

10.5 Explain how a telephone call is routed through a digital telephone exchange.

10.6 What is the integrated digital network (IDN)? Use a block diagram to show how local exchanges are connected to main switching centres and perhaps to other local exchanges. What factors are considered to decide whether two local exchanges are provided with direct trunk circuits?

10.7 List the different forms of bounded media that are used to carry signals in the PSTN. Optical fibre cable is increasingly used in preference to the other media; give some reasons for this. Why, so far, has optical fibre cable not been used to any great extent in the access network that connects the customer's telephone to the local exchange?

Chapter 11: Data communication

11.1 Give the reasons why digital signals cannot be directly transmitted over an analogue path such as the PSTN.

11.2 The data signal 11001101 is applied to a FSK modulator. Draw the waveform of the FSK signal, stating the frequencies used to represent both binary 1 and binary 0.

11.3 What is the function of a modem? Draw a circuit to show how two modems are employed to allow data to be transmitted over the PSTN.

11.4 Why is communications software needed when a computer is to transmit data over a circuit to a distant terminal or another computer? Would this be likely to be found in an integrated software package? An alternative to the PSTN for transmitting data is the PSDN. List the advantages to be gained from the use of the PSDN.

11.5 Explain the difference between a WAN and a LAN.

Draw two diagrams to show typical arrangements for each type of network.

11.6 A LAN makes use of 'servers'. Give some examples of different kinds of server and say when and why they would be employed.

Chapter 12: Radio systems

12.1 With the aid of a block diagram outline the operation of a commercial radio sound broadcast system.

12.2 Explain briefly the following terms used in conjunction with radio wave propagation: (a) ionosphere, (b) sky wave, and (c) wavelength.

12.3 Draw a sketch to show how signals in the VHF and UHF bands are propagated from one aerial to another. What is the name given to this mode of propagation and what is the maximum distance over which the wave can travel? What are the two ways in which the range can be increased.

12.4 Describe two methods of obtaining long-distance radio communication at frequencies in the LF and HF bands.

12.5 What is meant by an aerial? Will any conductor act as (a) a receiving aerial, (b) a transmitting aerial? Why do transmitting aerials work better at high frequencies than at low frequencies? An aerial works best if its length is about one-half wavelength long at its frequency of operation. Why, then, are some aerials mounted on the ground while others are mounted on top of masts?

12.6 State the function of each stage in a superheterodyne radio receiver. What would be the main difference between an amplitude modulation receiver and a frequency modulation receiver?

12.7 Cellular radio systems are increasingly used in modern communications. Explain the advantages of such a system and briefly outline its operation (a) when a mobile makes a call, and (b) when a mobile receives a call.

Chapter 13: Television

13.1 Explain why a televised image must be scanned at the transmitter and why the receiver must be in synchronism with this scan. What is interlaced scanning and why is it always used for television systems? How many lines are used in the UK system (a) in total, (b) in the visible picture?

13.2 Briefly explain the action of a CCD imager in producing a colour television picture signal. In what ways is a monochrome CCD imager different?

13.3 Draw a simple cross-sectional diagram of a monochrome television receiver tube and explain its operation.

13.4 Draw a simple cross-sectional diagram of a colour television receiver tube and explain its operation.

13.5 What are the primary colours in a television system and why do they vary from the primary colours in painting? What are the complementary colours in television? Explain the meanings of the terms 'hue', 'luminance' and 'saturation' that are used in colour television engineering. What are the colour-difference signals and why are they transmitted instead of the colour signals themselves?

13.6 Explain the differences that exist between the Teletext services Oracle and Ceefax and the viewdata system Prestel. How is a required page of data accessed in either system?

Answers to numerical exercises

6.5 Mark-space ratio = 0.5 (*Ans.*)
Duty factor = 0.33 (*Ans.*)
Mean voltage = 1 V (*Ans.*)
6.6 Resolution = $5/2^{12}$ = 5/4096 = 1.22 mV (*Ans.*)
8.1 09, 06 (*Ans.*)
9.2 1/500 = 2 ms (*Ans.*)

Index